THE DISPOSITION DILEMMA

Controlling the Release of Solid Materials from Nuclear Regulatory Commission-Licensed Facilities

Committee on Alternatives for Controlling the Release of Solid Materials
from Nuclear Regulatory Commission-Licensed Facilities

Board on Energy and Environmental Systems

Division on Engineering and Physical Sciences

National Research Council

NATIONAL ACADEMY PRESS
Washington, D.C.

National Academy Press • 2101 Constitution Avenue, N.W. • Washington, DC 20418

NOTICE: The project that is the subject of this report was approved by the Governing Board of the National Research Council, whose members are drawn from the councils of the National Academy of Sciences, the National Academy of Engineering, and the Institute of Medicine. The members of the committee responsible for the report were chosen for their special competences and with regard for appropriate balance.

This report and the study on which it is based were supported by Grant No. NRC-04-00-050. Any opinions, findings, conclusions, or recommendations expressed in this publication are those of the author(s) and do not necessarily reflect the view of the organizations or agencies that provided support for the project.

International Standard Book Number: 0-309-08417-2

COVER: Image adapted from a photograph of the decommissioned Big Rock Point nuclear plant near Charlevoix, Michigan, available at <http://www.consumersenergy.com/welcome.htm?./ocompany/index.asp?SS1ID=158>.

Available in limited supply from:
Board on Energy and Environmental
 Systems
National Research Council
2101 Constitution Avenue, N.W.
HA-270
Washington, DC 20418
202-334-3344

Additional copies available for sale from:
National Academy Press
2101 Constitution Avenue, N.W.
Box 285
Washington, DC 20055
800-624-6242 or 202-334-3313 (in the
 Washington metropolitan area)
http://www.nap.edu

Copyright 2002 by the National Academy of Sciences. All rights reserved.

Printed in the United States of America

THE NATIONAL ACADEMIES

National Academy of Sciences
National Academy of Engineering
Institute of Medicine
National Research Council

The **National Academy of Sciences** is a private, nonprofit, self-perpetuating society of distinguished scholars engaged in scientific and engineering research, dedicated to the furtherance of science and technology and to their use for the general welfare. Upon the authority of the charter granted to it by the Congress in 1863, the Academy has a mandate that requires it to advise the federal government on scientific and technical matters. Dr. Bruce M. Alberts is president of the National Academy of Sciences.

The **National Academy of Engineering** was established in 1964, under the charter of the National Academy of Sciences, as a parallel organization of outstanding engineers. It is autonomous in its administration and in the selection of its members, sharing with the National Academy of Sciences the responsibility for advising the federal government. The National Academy of Engineering also sponsors engineering programs aimed at meeting national needs, encourages education and research, and recognizes the superior achievements of engineers. Dr. Wm. A. Wulf is president of the National Academy of Engineering.

The **Institute of Medicine** was established in 1970 by the National Academy of Sciences to secure the services of eminent members of appropriate professions in the examination of policy matters pertaining to the health of the public. The Institute acts under the responsibility given to the National Academy of Sciences by its congressional charter to be an adviser to the federal government and, upon its own initiative, to identify issues of medical care, research, and education. Dr. Kenneth I. Shine is president of the Institute of Medicine.

The **National Research Council** was organized by the National Academy of Sciences in 1916 to associate the broad community of science and technology with the Academy's purposes of furthering knowledge and advising the federal government. Functioning in accordance with general policies determined by the Academy, the Council has become the principal operating agency of both the National Academy of Sciences and the National Academy of Engineering in providing services to the government, the public, and the scientific and engineering communities. The Council is administered jointly by both Academies and the Institute of Medicine. Dr. Bruce M. Alberts and Dr. Wm. A. Wulf are chairman and vice chairman, respectively, of the National Research Council.

COMMITTEE ON ALTERNATIVES FOR CONTROLLING THE RELEASE OF SOLID MATERIALS FROM NUCLEAR REGULATORY COMMISSION-LICENSED FACILITIES

RICHARD S. MAGEE, *Chair*, Carmagen Engineering Inc., Rockaway, New Jersey
DAVID E. ADELMAN, University of Arizona, Tucson
JAN BEYEA, Consulting in the Public Interest, Lambertville, New Jersey
JACK S. BRENIZER, JR., Pennsylvania State University, University Park
LYNDA L. BROTHERS, Sonnenschein, Nath & Rosenthal, San Francisco, California
ROBERT J. BUDNITZ, Future Resources Associates, Inc., Berkeley, California
GREGORY R. CHOPPIN, Florida State University, Tallahassee
MICHAEL CORRADINI, NAE,[1] University of Wisconsin, Madison
JAMES W. DALLY, NAE, University of Maryland, College Park
EDWARD R. EPP, Harvard University (retired), Cambridge, Massachusetts
ALVIN MUSHKATEL, Arizona State University, Tempe
REBECCA R. RUBIN, Businesses of Adams, Hargett and Riley Inc., Alexandria, Virginia
MICHAEL T. RYAN, Medical University of South Carolina, Charleston
RICHARD I. SMITH, Pacific Northwest National Laboratory (retired), Kennewick, Washington
DALE STEIN, NAE, Michigan Technological University (retired), Tucson, Arizona
DETLOF VON WINTERFELDT, University of Southern California, Los Angeles

Liaison from the Board on Energy and Environmental Systems

GERALD L. KULCINSKI, NAE, University of Wisconsin, Madison

Liaison from the Board on Radioactive Waste Management

ROBERT M. BERNERO, Nuclear Safety Consultant, Gaithersburg, Maryland

Project Staff

MARTIN OFFUTT, Study Director
JAMES ZUCCHETTO, Director, Board on Energy and Environmental Systems
RICK JOSTES, Program Officer, Board on Radiation Effects Research
PANOLA GOLSON, Project Assistant, BEES
SHANNA C. LIBERMAN, Project Assistant, BEES (until September 2001)

---

[1] NAE = member, National Academy of Engineering.

BOARD ON ENERGY AND ENVIRONMENTAL SYSTEMS

ROBERT L. HIRSCH, *Chair*, RAND, Arlington, Virginia
RICHARD E. BALZHISER, NAE,[1] Electric Power Research Institute, Inc. (retired), Menlo Park, California (term expired September 30, 2001)
DAVID L. BODDE, University of Missouri, Kansas City
PHILIP R. CLARK, NAE, GPU Nuclear Corporation (retired), Boonton, New Jersey
WILLIAM L. FISHER, NAE, University of Texas, Austin
CHRISTOPHER FLAVIN, Worldwatch Institute, Washington, D.C. (term expired August 31, 2001)
HAROLD FORSEN, NAE, Foreign Secretary, National Academy of Engineering, Washington, D.C.
ROBERT W. FRI, Resources for the Future, Washington, D.C.
WILLIAM FULKERSON, Oak Ridge National Laboratory (retired) and University of Tennessee, Knoxville
MARTHA A. KREBS, California Nanosystems Institute, Los Angeles, California
GERALD L. KULCINSKI, NAE, University of Wisconsin, Madison
JAMES MARKOWSKY, NAE, American Electric Power (retired), North Falmouth, Massachusetts
EDWARD S. RUBIN, Carnegie Mellon University, Pittsburgh, Pennsylvania
PHILIP R. SHARP, Harvard University, Cambridge, Massachusetts
ROBERT W. SHAW, JR., Aretê Corporation, Center Harbor, New Hampshire
JACK SIEGEL, Energy Resources International, Inc., Washington, D.C.
ROBERT SOCOLOW, Princeton University, Princeton, New Jersey
KATHLEEN C. TAYLOR, NAE, General Motors Corporation, Warren, Michigan
JACK WHITE, The Winslow Group, LLC, Fairfax, Virginia
JOHN J. WISE, NAE, Mobil Research and Development Company (retired), Princeton, New Jersey

Staff

JAMES ZUCCHETTO, Director
RICHARD CAMPBELL, Program Officer
ALAN CRANE, Program Officer
MARTIN OFFUTT, Program Officer
SUSANNA CLARENDON, Financial Associate
PANOLA GOLSON, Project Assistant
ANA-MARIA IGNAT, Project Assistant (until October 5, 2001)
SHANNA LIBERMAN, Project Assistant (until September 22, 2001)

---

[1] NAE = member, National Academy of Engineering.

Preface and Acknowledgments

Statutory responsibility for the protection of health and safety related to civilian nuclear facilities rests with the U.S. Nuclear Regulatory Commission (USNRC). The basic standard for protection against radiation is 10 CFR Part 20, which was first issued in final form by the Atomic Energy Commission in 1957 and was subject to a major revision that was finalized in 1991. Part 20 includes limits on quantities or amounts of radionuclides released in gaseous and liquid effluents below which the effluent would not be subject to further regulatory control, but it does not contain similar regulatory limits applicable to slightly radioactive solid material (SRSM). Absent such limits, the USNRC does have guidance documents regarding how slightly radioactive solid materials are cleared from regulatory control (a practice that licensees make use of routinely), and Section 2002 of Part 20 allows licensees to apply to the USNRC and its agreement states for clearance of solid materials on a case-by-case basis where the guidance documents do not apply. This policy issue could become increasingly important in the future as the eventual decommissioning of nuclear power plants generates large amounts of SRSM.

The USNRC has attempted without success to update and formalize its policies on disposition of SRSM. In 1990, it issued a policy, as directed by the Low Level Radioactive Waste Policy Amendments Act of 1985, that declared materials with low concentrations of radioactivity contamination "below regulatory concern" (BRC) and hence deregulated. However, Congress intervened to set aside the BRC policy in the Energy Policy Act of 1992, following the USNRC's own suspension of the policy. In 1999, the USNRC again examined the issue of disposition of SRSM and published a *Federal Register* notice examining several policy options. In neither case was the USNRC able to convince consumer and environmental groups that clearance of SRSM could be done safely or to convince some industry groups that clearance is desirable. In August 2000, the USNRC asked the National Research Council to form a committee to provide advice in a written report. The National Research Council established the Committee on Alternatives for Controlling the Release of Solid Materials from Nuclear Regulatory Commission-Licensed Facilities to address this task and recommend approaches for the clearance of solid materials from USNRC-licensed facilities (Appendix A contains biographical sketches of the committee members).

It became clear to the committee that radioactive waste is generated by many different industries and controlled by several government agencies under the terms of different regulations. This compounded the committee's task. During open information gathering sessions, the committee heard from stakeholders such as the U.S. Department of Energy (DOE) whose concerns focused on wastes that are not controlled by the USNRC; however, these stakeholders feared that any USNRC rulemaking or policy change might influence the disposition of these materials. Other large volumes of waste—e.g., naturally occurring radioactive materials (NORM) or technically enhanced NORM, which is known as TENORM—are not regulated under any specific federal statute. Finally, since the current case-by-case approach seems to be working, there is not a strong, unified impetus for change.

I wish to gratefully acknowledge the hard work of the committee members, who served as volunteers and who provided all the expertise necessary to carry out this difficult task. I am especially appreciative of the many hours they spent at the two writing sessions, which enabled us to complete the task on schedule. The assistance and contributions of the committee's two liaisons, Robert M. Bernero and Gerald L. Kulcinski, greatly enhanced the committee's efforts.

The presentations by U.S. Nuclear Regulatory Commission Chairman Richard A. Meserve; staff from the USNRC, the Environmental Protection Agency, and DOE; stakeholder organizations; nuclear industry representatives; representatives from the European Union and the International Atomic Energy Agency; and a host of other organizations, provided the committee with valuable information and insights into the issue of the disposition of SRSM from USNRC-licensed facilities. The contribution of these presenters is greatly appreciated (see Appendix B for a complete list of presentations).

Robert Meck at the USNRC was our principal point of contact; he ensured the constant flow of written information to the committee in response to our numerous questions and requests for additional information. Special thanks are owed to Al Johnson and Doug Jamieson, Duratek, Inc., for arrangements and a guided tour of Duratek's Bear Creek Operations (Oak Ridge, Tennessee) and Gallaher Road Facility (Kingston, Tennessee) and to Richard Grondin for a tour of the ATG, Inc., facility (Richland, Washington).

PREFACE AND ACKNOWLEDGMENTS   *ix*

A study such as this requires extensive support; we are all indebted to the National Research Council staff for their assistance. I would particularly like to acknowledge the close working relationship I had with the National Research Council study director, Martin Offutt, and the support I received from him. The logistic support that enabled us to concentrate on our task was ably provided by Shanna Liberman and Panola Golson. The efforts and contributions of the consulting technical writer, Robert Katt, greatly enhanced the clarity and sharpness of the report. The committee was also assisted by Alan Fellman, who provided technical consulting.

This report has been reviewed in draft form by individuals chosen for their diverse perspectives and technical expertise, in accordance with procedures approved by the National Research Council's Report Review Committee. The purpose of this independent review is to provide candid and critical comments that will assist the institution in making its published report as sound as possible and to ensure that the report meets institutional standards for objectivity, evidence, and responsiveness to the study charge. The review comments and draft manuscript remain confidential to protect the integrity of the deliberative process. We wish to thank the following individuals for their review of this report:

Vicki M. Bier, University of Wisconsin;
Philip R. Clark (NAE), General Public Utilities, Nuclear Corporation (retired);
Kenneth Eger, Jacobs Engineering Group;
Ann Fisher, Pennsylvania State University;
Gordon Geiger, University of Arizona;
Richard Guimond, Motorola, Inc.;
Ivan Itkin, former Director, Office of Civilian Radioactive Waste Management, U.S. Department of Energy;
David Lochbaum, Union of Concerned Scientists;
Claudio Pescatore, Organization for Economic Cooperation and Development, Nuclear Energy Agency;
John J. Taylor (NAE), Electric Power Research Institute, Inc.; and
Chris C. Whipple (NAE), Environ, Inc.

Although the reviewers listed above have provided many constructive comments and suggestions, they were not asked to endorse the conclusions or recommendations, nor did they see the final draft of the report before its release. The review of this report was overseen by Frank Parker (NAE) of Vanderbilt University. Appointed by the National Research Council, he was responsible for making sure that an independent examination of this report was carried out in accordance with institutional procedures and that all review comments were carefully considered. Responsibility for the final content of this report rests entirely with the authoring committee and the institution.

Richard S. Magee, *Chair*
Committee on Alternatives for
 Controlling the Release of Solid
 Materials from Nuclear Regulatory
 Commission-Licensed Facilities

Contents

EXECUTIVE SUMMARY   1

1   INTRODUCTION   13
    Historical Context, 14
    Radiation Protection Standards Developed by Organizations
        Other Than the USNRC, 20
    The U.S. and Global Contexts of Radioactive Waste Generation, 20
    Status of the Current USNRC Process for Clearing Solid Materials, 25
    The Study Task and Approach, 28

2   THE REGULATORY FRAMEWORK   33
    Mechanics of Existing and Former Standards Governing Releases
        of Radioactively Contaminated Material, 33
    Historical Evolution of the Regulatory Framework for Controlling
        Radioactively Contaminated Solid Materials, 39
    Comparative Assessment of Existing Regulations in the
        United States, 44
    Stakeholder Involvement, 52
    Findings, 53

3   ANTICIPATED INVENTORIES OF RADIOACTIVE OR   55
    RADIOACTIVELY CONTAMINATED MATERIALS
    Inventories of Contaminated Materials Arising from
        Decommissioning of USNRC-Licensed Facilities, 56

Inventories of Radioactive Waste from Other Licensed and
 Unlicensed Sources, 61
Findings, 71

4   PATHWAYS AND ESTIMATED COSTS FOR DISPOSITION   72
OF SLIGHTLY RADIOACTIVE MATERIAL
Disposition System Decisions, 73
Relative Costs for Disposition Alternatives, 75
Finding, 79

5   REVIEW OF METHODOLOGY FOR DOSE ANALYSIS   80
Key Technical Assessments of Annual Doses Associated with
 Clearance of Solid Materials, 81
USNRC Studies, 86
Environmental Protection Agency Documents on Dose Factors, 91
American National Standards Institute and Health Physics
 Society Standard N13.12-1999, 92
International Atomic Energy Agency Documents, 93
European Commission Documents, 95
Comparison of Clearance Studies, 96
Detailed Comments on NUREG-1640, 107
Findings, 112

6   MEASUREMENT ISSUES   115
Levels of Detectability, 117
Measurement Cost, 120
Current Measurement Practices of a Waste Broker, 122
The MARSSIM Methodology, 122
Findings, 124

7   INTERNATIONAL APPROACHES TO CLEARANCE   125
The Global Context, 125
Clearance Standards in the European Union, 131
Findings, 135

8   STAKEHOLDER REACTIONS AND INVOLVEMENT   136
Past USNRC Efforts at Stakeholder Involvement, 136
Risk Communication and Its Role in the Rulemaking Process, 144
Stakeholder Involvement: Methods and Successes, 147
Findings, 150

9	A FRAMEWORK AND PROCESS FOR DECISION MAKING Problems with the Current Approach, 151 The Decision-Making Process, 152 A Systematic Decision Framework, 154 Findings, 163	151
10	FINDINGS AND RECOMMENDATIONS Major Findings, 167 Recommendations, 171	166

REFERENCES 175

APPENDIXES
A	Biographical Sketches of Committee Members	183
B	Presentations and Committee Activities	192
C	Statement of Work	196
D	Standards (Limits) Proposed by Other Organizations	199
E	Radiation Measurement	212
F	Stakeholder Reactions to the USNRC Issues Paper	218
G	Acronyms and Glossary	230

Tables and Figures

TABLES

1-1 Average Annual Amounts of Ionizing Radiation to Which Individuals in the United States Are Exposed, 21
1-2 Common Sources of Radiation to Which the Public Is Exposed, 21
1-3 Risk Assessment Based on a Linear, No-Threshold Model with a Probability of Developing a Fatal Cancer of 5×10^{-2}/Sv (5×10^{-4}/rem), 30

3-1 Volume of Materials Arising from Power Reactor Decommissioning, 58
3-2 Weights of Slightly Radioactive Solid Material from Power Reactors, 59
3-3 Decommissioning Materials Inventory from the Population of U.S. Research Reactors, 62
3-4 Decommissioning Materials Inventory from the Population of U.S. Uranium Hexafluoride Conversion Plants, 64
3-5 Decommissioning Materials Inventory from the Population of U.S. Fuel Fabrication Plants, 64
3-6 Sites Containing Radioactively Contaminated Soils, 68
3-7 Sources, Quantities, and Concentrations of TENORM, 70

4-1 Approximate Costs for Disposal of Solid Material as Low-Level Radioactive Waste, 77
4-2 Estimated Costs for Alternative Dispositions of Slightly Radioactive Solid Material, 79

5-1 Technical Analyses Supporting Numerical Coefficients for Deriving Secondary Activity Standards from Primary Dose Standards, 84

5-2 NUREG-1640 Uncertainty Factors Averaged Across Radionuclides, 88

5-3 Comparison of Dose Factor Estimates Made to Support Clearance Proposals, 98

5-4 Ratio of NUREG-1640 Dose Factors to Other Estimates, Averaged Across Radionuclides, 100

6-1 Comparison of Derived Screening Levels and Laboratory Minimum Detectable Concentrations (MDCs) for Selected Radionuclides, 118

6-2 Detectability of Selected Radionuclides by Laboratory Analysis Relative to Derived Screening Level (DSL) from TSD 97, 119

6-3 Estimated Number of Analyzed Samples per Metric Ton of Waste at Breakeven Between Clearance and Low-Level Radioactive Waste Disposal, 121

7-1 International Clearance Status as of May 2001, 128

8-1 Matrix of Stakeholder Perspectives, 142

D-1 Exempt Quantities Established by Council Directive 96/29/EURATOM, 206

D-2 Derived USNRC Clearance Levels Assuming a 10 µSv/yr Primary Dose Standard (All Metals), 210

E-1 Radiation Sources and Their Activities, 217

FIGURES

ES-1 Time distribution for generation of slightly radioactive solid material from U.S. power reactor decommissionings, 5

3-1 Time distribution for generation of slightly radioactive solid material from U.S. power reactor decommissionings, 60

4-1 Decision points and disposition pathways, 73

5-1 Points at which technical information and judgments can inform rulemaking decisions related to clearance of slightly radioactive solid material, 82

5-2 Illustration of scenario pathways following SRSM clearance and hypothetical affected critical groups, 85

8-1 Dispute resolution techniques, 148

9-1 Decision impact matrix, 164

Executive Summary

The U.S. Nuclear Regulatory Commission (USNRC) and its predecessor, the U.S. Atomic Energy Commission (AEC), have attempted since the 1970s to give greater uniformity to the policy and regulatory framework that addresses the disposition of slightly radioactive solid material.[1] The issue remains unresolved and controversial. The USNRC has tried to issue policy statements and standards for the release of slightly radioactive solid material from regulatory control, while such material has been released and continues to be released under existing practices. In 1980 the USNRC proposed regulatory changes to deregulate contaminated metal alloys but withdrew them in 1986 and began work with the Environmental Protection Agency (EPA) to develop more broadly applicable federal guidance. In 1990 the USNRC issued a more sweeping policy, as directed by the Low Level Radioactive Waste Policy Amendments Act of 1985 (LLWPAA), declaring materials with low concentrations of radioactivity contamination "below regulatory concern" (BRC) and hence deregulated. Congress intervened to set aside the BRC policy in the Energy Policy Act of 1992, after the USNRC's own suspension of the policy. Subsequent attempts by USNRC staff to build consensus among stakeholder groups as a basis for future policy articulations were met by boycotts of stakeholder meetings, both in the immediate aftermath of the BRC policy and again in 1999 during public hear-

---

[1]The phrase "slightly radioactive solid material" is used to mean objects that contain radionuclides from licensed sources used or possessed by licensees of the USNRC and agreement states. These materials typically contain radionuclides at low concentrations, and by virtue of these low concentrations they can be considered for disposition as something other than low-level radioactive waste.

ings on a new examination of the disposition of such materials. The only USNRC standard addressing the disposition of slightly radioactive solid material is a guidance document published in June 1974 by the AEC, whose regulatory authority over civilian nuclear facilities the USNRC assumed upon its creation a few months later in January 1975.

In August 2000, with another examination of this issue under way, the USNRC requested that the National Research Council form a committee to provide advice in a written report. The National Research Council established the Committee on Alternatives for Controlling the Release of Solid Materials from Nuclear Regulatory Commission-Licensed Facilities to address this task. The committee's task involved evaluating and providing recommendations on the history of the technical bases and policies and precedents for managing slightly radioactive solid material from USNRC-licensed facilities; the sufficiency of technical bases needed to establish standards for release of solid materials from regulatory control ("clearance standards") and the adequacy of measurement technologies; the concerns of stakeholders and how the USNRC should incorporate them; and the efforts of international organizations on clearance standards. The committee was also asked to examine the current system for release of slightly radioactive solid material from regulatory control, to recommend whether the USNRC should continue to use this system and to recommend changes if appropriate. The committee's fact-finding process included two site visits to waste brokering facilities and nearly 40 invited presentations from the USNRC, the U.S. Department of Energy (DOE), and EPA staff; stakeholder organizations; nuclear industry organizations; and other interested parties.

A brief discussion is needed to describe the types of facilities regulated by the USNRC, the types of slightly radioactive solid material originating from these facilities, and which facilities are their principal source. As noted, the USNRC was split off from the AEC to regulate civilian nuclear facilities. It currently regulates 103 operating nuclear power reactors and 36 operating non-power reactors ("reactor licensees"), and approximately 5,000 specific materials licensees, which use or possess source, special nuclear, or byproduct material.[2] Some of the principal categories of facilities holding materials licenses include measuring system gauges and instruments (1,698 licenses), medical applications (1,556 licenses), and research and development facilities (474 licenses). Among

---

[2]Source material is uranium and thorium in natural isotopic ratios, or ores containing uranium and/or thorium above 0.05 percent by weight. Special nuclear material is plutonium, enriched uranium, and uranium-233. Byproduct material includes any radioactive material (except special nuclear material) yielded in or made radioactive by the process of nuclear fission. A second category of byproduct material is uranium mill tailings, added in 1978 ("11(e)(2) materials"). The foregoing definitions have been paraphrased from their original sources, the Atomic Energy Act and 10 CFR Part 20, to provide greater clarity. Those sources should be consulted with regard to the legal meaning and effect of these terms.

EXECUTIVE SUMMARY 3

the large facilities with materials licenses are 34 interim spent fuel storage facilities, 18 uranium mills, 7 uranium fuel fabrication plants, 2 uranium hexafluoride plants, and 2 uranium enrichment plants, all of which are components of the nuclear fuel cycle. The USNRC's agreement states[3] license roughly an additional 16,000 specific materials licensees.

Radioactive material is present at USNRC-licensed facilities in containment buildings; vehicles such as trucks and forklifts; and tools, piping, ductwork, or any other part of an object within a nuclear facility that has come into contact with radionuclides during normal operations or decommissioning. Surface contamination occurs when radioactive material remains on the surface of an otherwise uncontaminated object. Unlike volume contamination, it is sometimes easily removed using chemical or mechanical methods. Volume contamination occurs in a variety of ways, such as when radioactive material penetrates via cracks, pores, grain boundaries, or solid-state diffusion into an object or when incident neutrons activate (make radioactive) some of the atoms within an object. Volume contamination can also arise through mixing of radioactive material with solids such as soil. Objects having volume contamination are generally more difficult to decontaminate and are subject to a less-well-articulated system of standards for clearance from further regulatory control, as discussed below. The radiation emitted by radioactive material can have detrimental health effects on the various organs and tissues of the body, including induction of cancer. The unit of dose equivalent in the international system (SI), the sievert (Sv; equal to 100 rem) is used to indicate the biological effect of ionizing radiation and is used in setting radiation protection standards.

In conducting its study, the committee first examined the current system of standards, guidance, and practices used by the USNRC and agreement states to determine whether to release slightly radioactive solid material from further regulatory control under the Atomic Energy Act. The committee found that the current, workable system allows licensees to release material according to preestablished criteria but contains inconsistencies such that nuclear reactor licensees can release materials only if there is no detectable radioactivity[4] (above background levels), whereas materials licensees can do so if small detectable levels are found. The USNRC uses a guidance document for this latter purpose, Regulatory Guide

---

[3]Section 274 of the Atomic Energy Act (AEA) authorizes the Commission to enter into an effective agreement with the governor of a state to allow that state to assume the USNRC's authority to regulate certain types of materials licensees only. Reactor licensees remain the exclusive domain of the USNRC. Today there are 32 agreement states, which have implemented regulatory programs that are compatible with the USNRC's programs. The materials licensees that a state can regulate include those that use or possess material, byproduct material, or special nuclear material in quantities not sufficient to form a critical mass (e.g., less than 350 grams of uranium-235).

[4]Reactor licensees can, however, apply to USNRC for approval to release solid materials with small but detectable levels of radioactivity pursuant to Section 2002 of 10 CFR Part 20.

1.86, which includes a table of surface contamination limits that are technology based (measurement based) and not risk based (dose based). These limits are typically incorporated as license conditions or technical specifications in the case of materials licensees and subsequently used by the licensee to release material, whereas, as noted above, reactor licensees cannot release material if radioactivity is detected above natural background. No table of limits exists for volume contamination. Instead, the USNRC and its agreement states decide on a case-by-case basis whether release of volume-contaminated solid materials can occur. The committee found that licensees are currently submitting case-by-case applications at a rate that is being adequately managed by the USNRC and the agreement states.

Materials with levels of radioactivity not detectable above background radiation (with routine radiation measurements) are being released on a daily basis from nuclear power plants under a licensee arrangement with either the agreement states or the USNRC. In addition, some materials with volume contamination are being released on a case-by-case basis. The amount of these materials is not known, because there is no requirement to document the materials released. The annual dose equivalent resulting from these releases on a case-by-case basis has been estimated in draft NUREG-1640 at 10 µSv/yr (1 mrem/yr) or less for most of the radionuclides of interest.

The committee found that in future years the vast majority of slightly radioactive solid materials subject to the USNRC's system of clearance standards and practices will come from closing (decommissioning) nuclear power plants. Metal and concrete will constitute the greatest volume of slightly radioactive solid materials resulting from decommissioning. If power reactors are decommissioned on the schedule set by their current licenses, large quantities of metal and concrete waste will be generated during the next several decades, as shown in Figure ES-1. If licenses are extended for an additional 20 years, which seems probable for most facilities, the schedule shown in Figure ES-1 would be set back by as much as 20 years, with little material generated from decommissioning until after 2030.

The committee considered three general categories of options for disposition of slightly radioactive solid materials. *Clearance*[5] (unconditional—i.e., unrestricted—release) means that the material is handled as if it is no longer radioactive. Under this option, solid material (e.g., a tool) can be reused without restriction, recycled into a consumer product (e.g., a patio table), or disposed of in a landfill. (Classification of the waste as hazardous, for example, under the Resource Conservation and Recovery Act [RCRA], would depend on its other properties.) The committee found only limited support for clearance that allows

---

[5]Where the term clearance (i.e., no longer under regulatory control) appears, it is understood to mean unconditional clearance.

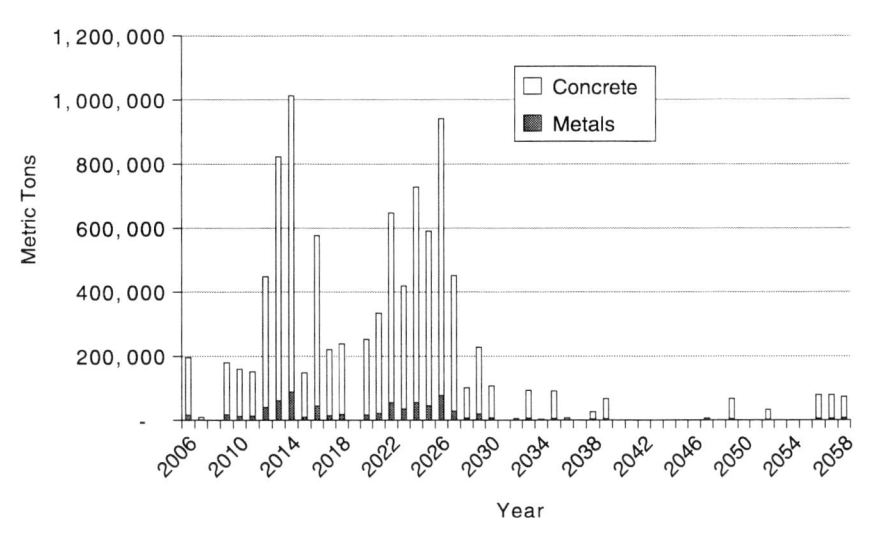

FIGURE ES-1 Time distribution for generation of slightly radioactive solid material from U.S. power reactor decommissionings. SOURCE: Adapted from SCA (2001).

slightly radioactive solid materials to enter commerce for unrestricted recycled use, no matter how restrictive the clearance standard might be. No support for this option exists in the steel and concrete industries.

Conditional clearance (i.e., restricted release from regulatory control) means that material must be used in a specified application and subject to continuing regulatory control until specific conditions are met. For example, slightly radioactive metal released under a conditional clearance standard might be melted into shielding blocks for use at DOE nuclear facilities but could be subject to controls in the process. Other examples might include slightly radioactive concrete that must be disposed of in a Subtitle D landfill or concrete that is released for use in the rubble base for roads. Conditionally cleared material would not be released for use in general commerce.

No release (from regulatory control) means that the slightly radioactive solid material, once it leaves the originating facility, must be sent to a facility licensed to accept radioactive solid material for storage or disposal. Under this option, the slightly radioactive solid material remains under a USNRC or agreement state license continuously.[6] Under current conditions, slightly radioactive solid material would be sent to either Envirocare of Utah or one of two disposal facilities licensed to accept all types of low-level radioactive waste (LLRW)—Barnwell,

---

[6]Until the expiration of postclosure monitoring requirements.

South Carolina, or U.S. Ecology in Richland, Washington—in accordance with each facility's licenses and permits. Each general disposition option—clearance, conditional clearance, and no release—has minor variants and regulatory complexities, which are discussed in this report.

Each disposition option has economic implications due to associated pricing and handling, regulation, and disposal. (For estimation purposes, only concrete and metal are considered.) If the material is disposed of as radioactive waste, as in the case of "no release," then the disposal fee charged by the facility could range from $3,120 (U.S. Ecology) to $16,800 (Barnwell) per cubic meter in the two licensed commercial low-level radioactive waste disposal facilities. The cost to dispose of slightly radioactive metal from all U.S. power reactors would range from $1.6 billion to $8.8 billion, depending on whether U.S. Ecology or Barnwell is used, respectively.[7] For slightly radioactive concrete, the committee estimates disposal at Envirocare of Utah could be accomplished at a cost roughly one eighth that of U.S. Ecology, giving a total cost for all concrete from U.S. power reactors of $2.9 billion.[8] The total cost to dispose of all slightly radioactive solid material—metal and concrete—from U.S. power reactors under the no-release option is thus estimated at between $4.5 billion and $11.7 billion. Less costly disposal is possible if the slightly radioactive solid material meets the terms of conditional clearance and can be sent to a landfill. Then disposal can be accomplished at a disposal fee of approximately $30 per metric ton for a Subtitle D landfill (municipal waste) and approximately $110 per metric ton for a RCRA Subtitle C landfill (hazardous waste). Disposal of all the slightly radioactive solid material anticipated from U.S. power reactors could cost $0.3 billion in Subtitle D landfills and $1 billion in Subtitle C landfills. Clearance of all this material could allow the option of recycle or reuse for some of the material, as appropriate, and would avoid essentially all disposal costs. These estimates are shown to illustrate the relative costs of the different clearance policy options; it should be emphasized, however, that the cost of disposal of slightly radioactive solid materials may in the future be subject to factors that the committee is not able to foresee or take into account. For example, the committee has not considered energy deregulation or the impact on ratepayers caused by any changes that may be made to clearance rules.

Licensees will base decisions on which disposition option is appropriate for

---

[7]Envirocare of Utah is licensed to accept bulk metal for disposal but does not publish pricing information and determines prices on a case-by-case basis. The committee was not able to find data on such prices for disposal of bulk metals at Envirocare, so it has not estimated the costs of disposal of metal from U.S. power reactors.

[8]Envirocare of Utah charges the U.S. Army Corps of Engineers $298 per cubic yard ($388 per cubic meter) for disposal of high-volume, slightly radioactive concrete debris, which is classified as pre-1978 uranium mill tailings by the USNRC.

EXECUTIVE SUMMARY

a particular quantity of slightly radioactive solid material, in part, on measurements of the amounts of radioactive materials present. Measurement of the amount of radioactive material in a solid matrix is a function of instrument characteristics, background radiation levels, and source characteristics. If the sampling and analysis costs are too high, it may be more cost-effective to dispose of the material at a facility licensed to accept low-level radioactive waste rather than demonstrate compliance with a clearance standard to allow landfill disposal. For screening-level concentrations and surface contaminations calculated from dose levels greater than or equal to 10 µSv/yr (1 mrem/yr), for a defined exposure scenario, detection is possible in a laboratory setting for a majority of radionuclides under most practical conditions at reasonable costs. Using field measurements, a more rapid fall-off of detectability is observed at more stringent radiation protection levels, with 31 of 40 key radionuclides detectable at 10 µSv/yr (1 mrem/yr) and 11 of 40 detectable at 1 µSv/yr (0.1 mrem/yr).

The committee evaluated technical analyses of the estimated doses of the final disposition of slightly radioactive solid materials. These analyses were conducted by federal agencies and international organizations, including the International Atomic Energy Agency (IAEA), the European Commission, and other groups. The committee paid particular attention to a draft USNRC document, NUREG-1640, which was developed to support its most recent evaluation of the clearance standard issue. The committee concluded that of the various reports considered, draft NUREG-1640 provided a *conceptual framework*, particularly with regard to incorporating formal uncertainty, that best represents the current state of the art in risk assessment. The committee did find limitations in the report, including its lack of applicability to scenarios of conditional clearance (e.g., landfill disposal), lack of consideration of multiple exposure pathways, and lack of consideration of human error[9] and its possible effect on dose factor prediction. Draft NUREG-1640 has also been clouded by questions of contractor conflict of interest.

To determine if numerical values in the report had been affected by considerations other than science, the committee checked a sample of dose factor analyses and found them reasonable. Once all of the dose factors are checked as the committee recommends and other limitations in draft NUREG-1640 have been resolved—either in the final version of the report or in follow-up reports—the resulting dose factors can be multiplied by appropriate dose-risk coefficients to provide estimates of the risks of releasing individual radionuclides at any hypothetical concentration. The USNRC will then have a sound basis for considering

---

[9]Human error is used here to mean the violation of scenario assumptions at some infrequent, but nonzero, rate. Categories of relevant human error include mistakes in properly labeling material, mistakes in measurement, or failure to properly decontaminate loose material as assumed in dose factor estimates.

> **BOX ES-1**
> **Policy Alternatives for Releasing**
> **Slightly Radioactive Solid Material**
>
> *Case-by-Case Approach*
>
> - Current approach: USNRC or agreement state approves specific license conditions
> - Additional criteria for volume contamination
> - Restrictions on reuse (see examples below, under "conditional clearance")
>
> *Clearance Standard*
>
> - Dose based (based on risk to an individual or population caused by exposure to radiation)
> - Source based (based on surface or volume radioactivity concentration of the contaminated solid material)
>
> *Conditional Clearance Standard*
>
> - Dose based (based on risk to an individual or population caused by exposure to radiation)
>   —Beneficial reuse in controlled environments (e.g., metal for shield blocks in USNRC-licensed or DOE facilities)
>   —Limited reuse for low-exposure scenarios (e.g., concrete rubble base for roads)
>   —Landfill disposal
> - Source based (based on surface or volume radioactivity concentration of the contaminated solid material)
>   —Beneficial reuse in controlled environments (e.g., metal for shield blocks in USNRC-licensed or DOE facilities)
>   —Limited reuse for low-exposure scenarios (e.g., concrete rubble base for roads)
>   —Landfill disposal
>
> *No Release*
>
> - All slightly radioactive solid material is disposed of at licensed LLRW sites.

the total risks associated with any proposed clearance standards and for assessing the uncertainty attached to dose estimates. The committee does not believe it is necessary from a scientific perspective for the USNRC to start all over again.

The committee reviewed efforts by other countries and international organizations to set clearance standards. The European Union has issued a safety directive containing tables derived using a scenario assessment process against which slightly radioactive solid materials can be evaluated for possible clearance from further regulatory control. Member nations of the European Union are in the process of implementing this directive.

The issue of releasing radioactive materials from further regulatory control, like the issue of nuclear power in general, has received significant stakeholder input. The committee found that in the past, the USNRC failed to convince any environmental and consumer advocacy groups that the clearance of slightly radioactive solid material can be conducted safely and failed to convince certain industry groups that such clearance is desirable. Most of the issues and concerns expressed today by many consumer advocacy and environmental groups and some industry groups are the same as were expressed during the controversy over the BRC policy in 1990. Furthermore, a legacy of distrust of the USNRC has developed among many of the environmental stakeholder groups, resulting from their experience with the BRC policy, the enhanced participatory rulemaking on license termination ("decommissioning rule"), and the USNRC's 1999 issues paper, published in the *Federal Register* on June 30, 1999, on the clearance standards. Reestablishing trust will require concerted and sustained effort by the USNRC.

The committee developed a series of policy alternatives to the current system, detailed in Box ES-1. The committee found that there is time for the USNRC to move forward and select from among the alternatives, since no evidence was found that the problems associated with the current case-by-case approach required its immediate replacement. The committee does not recommend any one particular alternative. Instead, it emphasizes the need for the USNRC to undertake with deliberate speed and a broad range of stakeholder involvement a detailed and thorough analysis and evaluation of various alternative approaches that proceeds from logical starting points based on a sound technical foundation. Should the USNRC choose to develop new regulations for clearance, it has to consider that any action it takes may have implications for the management of materials—e.g., technologically enhanced naturally occurring radioactive materials (TENORM)—that are not currently regulated by the USNRC, DOE, or agreement states.

Considerations include effects on public health, costs and benefits, consistency with existing national and international analysis, practice and legal authority, and public perceptions and acceptance.

RECOMMENDATIONS

In developing its recommendations the committee was guided by two overarching, compelling findings:

1. The current approach to clearance decisions is workable and is sufficiently protective of public health that it does not need immediate revamping. However, the current approach, among other shortcomings, is inconsistently applied, is not explicitly risk based, and has no specific standards in guidance or regulations for clearance of volume-contaminated slightly radioactive solid material. Therefore, the committee believes that the USNRC should move ahead without delay and start a process of evaluating alternatives to the current system and its shortcomings.
2. Broad stakeholder involvement and participation in the USNRC's decision-making process on the range of alternative approaches is critical as the USNRC moves forward. The likelihood of acceptance of a USNRC decision greatly increases when the process (1) engages all responsible stakeholder representatives and viewpoints, (2) is perceived by participants as fair and open, (3) addresses all the advantages and disadvantages of the alternative approaches in an even-handed way, and (4) is open to a broad and creative range of alternatives. Thus, it is essential that the USNRC focus on the process and not prescribe an outcome. The outcome, an approach to disposition of slightly radioactive solid material, must evolve from the process.

While the committee did not want to prescribe the outcome of the decision process, it has made several specific recommendations, conditional on the process arriving at certain decision points. For example, if the USNRC contemplates clearance or conditional clearance standards, the committee recommends that these standards be dose based. The committee also recognized that significant national and international efforts have been completed, or are near completion, that provide a solid foundation for the USNRC to move forward. The committee has recommended the foundation from which to begin the process. Thus, the USNRC should be able to proceed expeditiously with a broad-based stakeholder participatory decision-making process.

Recommendation 1. The USNRC should devise a new decision framework that would develop, analyze, and evaluate a broader range of alternative approaches to the disposition of slightly radioactive solid material. At a minimum, these alternatives should include the current case-by-case approach, clearance, conditional clearance, and no release.

Recommendation 2. The USNRC's decision-making process on the range of alternative approaches to the disposition of slightly radioactive solid material should be integrated with a broad-based stakeholder participatory decision-making process. Elements of this process should include the following:

- The willingness and commitment of the USNRC to establish and maintain a meaningful and open dialogue with a wide range of stakeholders regarding the disposition of slightly radioactive solid material;
- An ad hoc broad-based advisory board that would advise the USNRC in its consideration of approaches to the disposition of slightly radioactive solid material. The advisory board would also suggest additional stakeholder involvement mechanisms that the USNRC could use in the decision process (for example, establishing a National Environmental Policy Act process; alternative dispute resolution; and partnering, arbitration, mediation, or a combination of such methods); and
- Assistance obtained by the USNRC as needed from outside experts in order to (1) assist its efforts to establish the ad hoc stakeholder advisory board and to facilitate dialogue between the USNRC and stakeholder participants in the decision-making process and (2) assess, evaluate, and perhaps conduct portions of the USNRC stakeholder involvement program and make recommendations as appropriate.

Recommendation 3. The USNRC should adopt an overarching policy statement describing the principles governing the management and disposition of slightly radioactive solid material. A review and discussion of the IAEA policy statement *Principles for the Exemption of Radiation Sources and Practices from Regulatory Control* (Safety Series No. 89, IAEA Safety Guidelines, Vienna, 1988) with a broad-based stakeholder group would provide a good starting point in developing a policy statement that would provide a foundation for evaluation of alternative approaches to disposition of slightly radioactive solid material.

Recommendation 4. When considering either clearance or conditional clearance, a dose-based standard should be employed as the primary standard. To employ a dose-based standard, it is necessary to consider a wide range of scenarios that encompass the people likely to be exposed to slightly radioactive solid material. From these people, a critical group is selected and secondary standards (based on dose factors) are derived. These secondary standards are used to limit the radioactivity in materials being considered for release or conditional release.

The USNRC should also consider the pros and cons of the establishment of a separate collective dose standard.

Recommendation 5. An individual dose standard of 10 μSv/yr (1 mrem/yr) provides a reasonable starting point for the process of considering options for a dose-based standard for clearance or conditional clearance of slightly radioactive solid material. This starting point is appropriate for the following reasons:

- A dose of 10 μSv/yr (1 mrem/yr) is a small fraction (less than 0.5 percent) of the radiation received each year from natural background sources.

- A dose of 10 µSv/yr (1 mrem/yr) is significantly less than the amount of radiation that we receive from our own body due to radioactive potassium (one contributor to background radiation) and other elements and to routine medical procedures that involve ionizing radiation.
- A dose of 10 µSv/yr (1 mrem/yr) over a 70-year lifetime equates to an estimated increase of 3.5×10^{-5} in the lifetime cancer risk, which falls within the range of acceptable lifetime risks of 5×10^{-4} to 10^{-6} used in developing health-based standards for exposure to radiation (other than for radon) in the United States.
- Radiation measurement technologies are available at a reasonable cost to detect radioactivity at concentrations derived from this dose standard.
- This dose standard is widely accepted by recognized national and international organizations.

The final selection of an individual dose standard should nonetheless be a policy choice, albeit one informed by the above considerations.

Recommendation 6. For any dose-based alternative approach to disposition of slightly radioactive solid materials, the USNRC should use the *conceptual framework* of draft NUREG-1640 to assess dose implications. To use the actual results of NUREG-1640 in the decision framework discussed in Recommendations 1 and 2, the USNRC must first establish confidence in the numerical values, expand the scope of applicability, and overcome certain limitations in draft NUREG-1640. At a minimum, the following specific actions are required:

- Review the choice of parameter distributions used in the dose modeling, as well as the characteristic values chosen for each parameter distribution.
- Develop complete scenarios and dose factors for conditional clearance options.
- Provide sufficient information to enable calculation of collective doses to support Recommendation 4.
- Expand the current set of scenarios used to compute dose factors to include (1) human error and (2) multiple exposure pathways.

The USNRC should use an independent group of experts to provide peer review of these activities.

Recommendation 7. The USNRC should continue to review, assess, and participate in the ongoing international effort to manage the disposition of slightly radioactive solid material. The USNRC should also develop a rationale for consistency between secondary dose standards that may be adopted by the United States and other countries. However, the USNRC should ensure that the technical basis for secondary dose standards is not adjusted for consistency unless these adjustments are supported by scientific evidence.

1

Introduction

The charge to the committee was to study possible approaches for releasing slightly radioactive solid material from U.S. Nuclear Regulatory Commission (USNRC)-licensed facilities. Accordingly, the analyses in the first nine chapters and the recommendations in Chapter 10 pertain primarily to slightly radioactive solid materials currently under the regulatory control of the USNRC or agreement states.[1]

The term "slightly radioactive solid material" (SRSM) refers to material that contains radionuclides from licensed sources used or possessed by licensees of the USNRC and agreement states. These materials typically contain low concentrations of radionuclides and, by virtue of these low concentrations, can be considered for disposition as something other than low-level radioactive waste (LLRW).[2]

---

[1] Section 274 of the Atomic Energy Act (AEA) authorizes the Commission to enter into an effective agreement with the governor of a state to allow that state to assume the USNRC's authority to regulate certain types of materials licensees only. Reactor licensees remain the exclusive domain of the USNRC. Today there are 32 agreement states, which have implemented regulatory programs that are compatible with the USNRC's programs. The materials licensees that a state can regulate include those that use or possess source material, byproduct material, or special nuclear material in quantities not sufficient to form a critical mass (e.g., less than 350 grams of uranium-235).

[2] LLRW is waste that contains concentrations of radioactive materials that are regulated under 10 CFR Part 61. There is no low-end cutoff for the concentrations of radioactive materials regulated as LLRW.

This chapter begins with the historical context for current USNRC regulations pertinent to the release of solid materials from licensed facilities. Next is a review of approaches used by other agencies for release (removal) of radioactive materials from regulatory control and a summary of the current process by which the USNRC decides on the release of solid materials using a case-by-case approach. The chapter concludes with a summary of the committee's task (the full text of the statement of work can be found in Appendix C) and a synopsis of the role each chapter plays in fulfilling that task.

HISTORICAL CONTEXT

The USNRC's basic standards for protection against radiation are set forth in 10 CFR Part 20,[3] a regulation intended "... to control the receipt, possession, use, transfer, and disposal of licensed material...." This regulation was first issued as a final regulation by the Atomic Energy Commission (AEC) in 1957 and was used for many years with minor amendments. The 1957 version of 10 CFR Part 20 contains a short section on waste disposal that provides the basis for case-by-case review of disposal procedures not covered within the two succeeding sections that deal with disposal of tritium and carbon-14 in sewerage systems or in soil. The 1957 regulation did not include criteria specifying an amount or concentration of a radionuclide in a solid material,[4] below which the solid material would be exempt from regulatory control or conditional clearance (Box 1-1).[5] However, pursuant to Section 2002 of 10 CFR Part 20, added in a later revision of the regulation, the USNRC and agreement states evaluate requests by licensees for permission to release solid materials on a case-by-case basis, using existing regulatory guidance.[6] The situation for gaseous and liquid materials is different,

---

[3]References to the United States Code of Federal Regulations (CFR) will be given using the conventional format with the code title (here, Title 10) followed by the acronym CFR and the part or chapter number(s).

[4]For two radionuclides, in one specific application, Part 20 does contain release criteria for solid materials. These criteria allow disposal of volume-contaminated animal tissue containing less than 1.85 kBq/g of ^{3}H or ^{14}C as if it were not radioactive.

[5]The definitions of terms related to release of materials from regulatory control are presented in Box 1-1. The committee notes much confusion about the common usage of terms in discussion of the release of radioactive materials. Without necessarily affirming this approach, the committee decided to use the terms as defined in the American National Standards Institute-Health Physics Society (ANSI/HPS, 1999) Standard N13.12-1999.

[6]The 1957 issue of Part 20 had a short section on waste disposal that included Part 20.302, "Method for obtaining approval of proposed disposal procedures," the basis for case-by-case review of disposal procedures not authorized by the two succeeding sections on disposal in sewerage systems or in soil. The original Part 20 gave general requirements for waste disposal of byproduct material. The 1957 standard did not include any criteria for a floor to the amount or concentration of controlled radionuclides, which criteria might be used as the basis for exemption of waste from regulatory control.

BOX 1-1
Definition of Selected Terms Related to Clearance of Materials from Nuclear Facilities

Background radiation. Natural radiation or radioactive material in the environment, including primordial radionuclides, cosmogenic radionuclides, and cosmic radiation. Primordial radionuclides belong to one of the three radioactive decay series headed by (1) ^{238}U, ^{235}U, and ^{232}Th; (2) ^{40}K; or (3) ^{87}Rb. Cosmogenic radionuclides are produced by collision of cosmic nucleons with atoms in the atmosphere or in the earth, including ^{14}C, ^{3}H, ^{7}Be, and ^{22}Na. Cosmic radiation comes from the secondary particles, mostly high-energy muons and electrons, produced by interactions between the earth's atmosphere and charged particles, primarily protons, from extraterrestrial sources. Naturally occurring radioactive material that has been technologically enhanced is not considered background for the purposes of the American National Standards Institute–Health Physics Society standard.

Clearance. The removal of items or materials that contain residual levels of radioactive materials employed within authorized practices from any further control of any kind.

Conditional clearance. The act of removing items or materials that contain residual levels of radioactive materials from regulatory control albeit with restrictions on the further use of the items or materials.

Exclusion. The designation by a regulatory authority that the magnitude or likelihood of an exposure is essentially not amenable to control through requirements of a standard and that such an exposure is outside the scope of standards (e.g., exposures from ^{40}K in the body, from cosmic radiation at the surface of the earth, and from unmodified concentrations of radionuclides in most raw materials).

Exemption. The designation by a regulatory authority that specified uses of radioactive materials or sources of radiation are not subject to regulatory control because the radiation risks to individuals and the collective radiological impact are sufficiently low.

Surface contamination. Radioactive contamination residing on or near the surface of an item. This contamination can be adequately quantified in units of activity per unit area. When an item has been exposed to neutrons (including structural components and shielding at nuclear reactors), or when an item could have cracks or interior surfaces allowing the distribution of radioactive contamination within the interior matrix, it is considered to be a volume contamination source.

Volume contamination. Radioactive contamination residing in or throughout the volume of an item. Volume contamination can result from neutron activation or from the penetration of radioactive contamination into cracks or interior surfaces within the matrix of an item. (Volume contamination can also occur due to solid-state diffusion.)

SOURCE: Adapted from ANSI/HPS (1999).

and Part 20 does set limits on the amount or concentration of a radionuclide in such materials that may be released to the environment from a nuclear facility. These concentration limits, which have been set for essentially all radionuclides of concern (numbering in the hundreds), are based on calculated dose to the general public. Volume-contaminated facility structures and soils that remain at decommissioned sites are regulated under Part 20, Subpart E, which establishes criteria for unrestricted use.

In June 1974 the AEC issued Regulatory Guide 1.86, *Termination of Operating Licenses for Nuclear Reactors* (AEC, 1974). This guide provides four alternatives for retiring a reactor facility at the end of its operational life. After the facility or equipment has been decontaminated and if the residual surface radiation levels do not exceed the limits stated in Table I of Regulatory Guide 1.86, the licensee may release the equipment or the USNRC may authorize termination of the facility license. Ever since the guide was issued, Table I has been used as a basis for releasing surface-contaminated material from further regulatory control when appropriate—for example, when incorporated into the conditions of a license.

In 1991 the USNRC, as the successor agency to the AEC for regulating nuclear facilities, issued a major revision to 10 CFR Part 20. The stated purpose of this revision was ". . . to modify the [US]NRC's radiation protection standards to reflect developments in the principles and scientific knowledge underlying radiation protection that have occurred since Part 20 was originally issued more than 30 years ago" (USNRC, 1991c). The revision also discusses its relationship to the recommendations of the International Commission on Radiological Protection (ICRP) and its U.S. counterpart, the National Council on Radiation Protection and Measurements (NCRP). Information was provided about the revisions to the Federal Radiation Protection Guidance on Occupational Exposure—which incorporate the philosophy and methodology of ICRP Parts 26 and 30—and the recently issued revisions in NCRP Report 91 (NCRP, 1987c) of the 1971 recommendations on radiation protection limits. The recommendation in NCRP Report 91 for a negligible individual risk level of 1 mrem/yr (0.01 mSv/yr) was recognized but not adopted by the USNRC for procedural reasons (NCRP Report 91 was issued after the proposed Part 20 rule, and there had been no opportunity for public comment). Box 1-2 contains definitions of the units of measurement used in this report.

The 1991 revision to 10 CFR Part 20 included other references on radiation protection, including a 1988 report of the United Nations Scientific Committee on the Effects of Atomic Radiation (UNSCEAR, 1988), reports by committees of the National Research Council (NRC, 1990) on the Biological Effects of Ionizing Radiation (BEIR), and the 1990 recommendations of the ICRP (ICRP, 1990). The 1991 revision also included allowable limits on the radiation dose that an individual could receive from exposure to radioactive materials (*dose limits*) and the concentration limits for radioisotopes released in gaseous or liquid effluents.

Even before the 1991 revision to Part 20 was issued, the USNRC, international governments, and non-U.S. agencies had agreed on a principal dose limit for members of the public of 100 mrem/yr, rather than the old limit of 500 mrem/yr. Although the USNRC has agreed to this dose limit set by the ICRP, the U.S. Environmental Protection Agency (EPA) has not yet done so (ICRP, 1985; USNRC, 1991c). This exposure limit was chosen with the recognition that the average exposure due to natural background radiation had been estimated at 240 mrem/yr by UNSCEAR and 300 mrem/yr by NCRP (UNSCEAR, 1982; NCRP, 1987a). In revising Part 20, the USNRC recognized that "when application of the dose limits is combined with the principle of keeping all radiation exposures 'as low as is reasonably achievable' [ALARA] the degree of protection could be significantly greater than from relying upon the dose limits alone." Part 20 as revised sets dose limits compatible with ALARA.

In issuing a standard for the uranium fuel cycle, the EPA allocated a public exposure limit of 25 mrem/yr, whole-body effective dose,[7] to the fuel cycle (40 CFR Part 190). All of the regulatory bodies use these exposure limits in the context of the three principles of radiation protection:

1. Justification of a practice;
2. Optimization (USNRC makes explicit use of ALARA—exposures held as low as is reasonably achievable);[8] and
3. Limitation of individual risk through exposure limits.

In the text of the revised 10 CFR Part 20, the USNRC recognized that the ALARA standard for reactor effluent releases, combined with the EPA fuel cycle standard, in effect set a limit on exposure of the general public to radioactive effluents that was only a few percent of the USNRC dose limit of 100 mrem/yr.

Optimization through an ALARA standard is central to the USNRC's radiation protection strategy. The objective is not merely to meet the dose limit but to go below it as far as is reasonably achievable. One way to address the possibility of doses to some members of the general public arising from multiple exposures to different clearance practices is to rely on the unquantified margin induced by

---

[7]Also included in this standard were limits of 75 mrem effective dose to the thyroid and 25 mrem effective dose to any one organ. The system of effective dose predates the system of dose equivalent now in widespread use, and the two are not directly comparable. The EPA has equated 25 mrem/yr whole-body effective dose to 15 mrem/yr dose equivalent (58 Federal Register 66398-66416; December 20, 1993).

[8]The EPA does not apply the optimization principle in the same way that the USNRC does. The EPA implements this principle broadly within its multistatute mission.

> **BOX 1-2**
> **Units of Measurement for Radiation Dose**
>
> This report deals with ionizing radiation. There are two types of directly ionizing radiation. X rays and gamma rays have the same characteristics and properties; they are both electromagnetic radiation and differ only in their source. X rays are emitted from electrical devices, where they are produced when electrons decelerate, or from atoms when energetic electrons move to vacancies in lower orbital shells. Gamma rays are emitted from nuclei of atoms during radioactive decay. The other type of directly ionizing radiation consists of highly energetic subatomic particles carrying a net electric charge, including electrons, protons, and alpha particles.
>
> Neutrons, which are uncharged particles, give up their energy by colliding with atomic nuclei, particularly so when colliding with particles of comparable mass. Neutrons are emitted from atomic nuclei when some radioactive materials undergo fission, thereby splitting into smaller atoms.
>
> Electrons are small negatively charged particles found in all atoms. When radioactive materials decay, the electrons that are emitted from decaying nuclei are known as beta particles.
>
> Alpha particles, which consist of two protons and two neutrons, are identical to the nucleus of a helium atom. Alpha particles are commonly emitted when higher-mass radionuclides such as uranium or radium decay.
>
> The amount of ionizing radiation to which an organism is exposed, which is usually called the radiation dose, can be measured in terms of the energy absorbed in matter. Regardless of the type of radiation, the energy of absorbed ionizing radiation is measured in units of rads. When the amount of radiation is small, the unit used is the millirad (1,000 millirads = 1 rad). One rad is equal to an absorbed dose of 100 ergs per gram of absorbing matter, or 0.01 joule/kg.

an ALARA standard.[9] Another approach is to allocate a fractional part of the dose limit to a practice, as EPA did in the facility standard for the uranium fuel cycle.

Along with establishing a dose limit for individual members of the public, the Part 20 revision for decommissioning allocated a significant fraction of the general limit to individual facilities. This approach appears reasonable, since it is difficult to envision that more than a few facilities would simultaneously be the source of significant exposure to any member of the public because the facilities are at fixed sites.

The USNRC has tried previously to set standards for release of SRSM from regulatory control. A proposed rule (45 Federal Register 70874; October 27,

---

[9]The USNRC regularly applies ALARA with protection limits but recognizes that the margin induced by ALARA can vary widely from case to case—for example, the contrast in site decommissioning between users of sealed sources and users of unsealed quantities of radioactive materials (59 Federal Register 43208). Also, the ALARA concept would become irrelevant at the proposed de minimis levels of clearance standards.

INTRODUCTION

> Doses of ionizing radiation with the same energy but involving different particles do not produce equal biological effects. In general, X rays and gamma rays are less damaging than alpha particles, neutrons, and protons. To account for these differences, a derivative unit is generally used. The unit customarily used in the United States is the rem. A radiation dose in rems is equal to the dose in rads multiplied by a quality factor to allow for the damage effectiveness of the type of particle involved. X-rays and gamma rays have been assigned a quality factor of 1. The electrically charged subatomic particles have quality factors greater than 1.
>
> There are two systems of units employed for measuring radiation doses, the U.S. Customary and the SI systems. The gray (Gy) is the SI unit of absorbed dose. One gray is equal to an absorbed dose of 1 joule/kg, or 100 rads. The SI unit for dose equivalent is the sievert (Sv). The dose equivalent in sieverts is equal to the absorbed dose in grays multiplied by the same quality factor used to convert from rads to rems. The conversion factor between the two dose equivalent units is 1 Sv = 100 rem, or 1 μSv = 100 mrem, or 10 mSv = 1 mrem.
>
> The becquerel (Bq) is the SI unit for the amount of radioactivity in a substance, measured by the rate of decay of radionuclides in the material. One becquerel is equal to one disintegration per second. Another unit used for the rate of decay is disintegrations per minute (dpm), and 60 dpm equals 1 Bq. The curie (Ci) is the U.S. Customary unit of measure for the amount of radioactivity as indicated by the rate of decay of a radioactive material; 1 Ci equals 3.7×10^{10} Bq, or 2.22×10^{12} dpm.
>
> Individual dose is the dose received by the exposed individuals in the critical group of the exposed segment of the population.
>
> Collective dose is the sum of the doses received in a given period of time by a specified population from exposure to a specified source of radiation.

1980) to exempt residual levels of radionuclides in smelted alloys from licensing was withdrawn in 1986 (51 Federal Register 8842; March 14, 1986). A more sweeping policy issued by the USNRC, as directed by the Low Level Radioactive Waste Policy Amendments Act of 1985 (LLWPAA), declared materials with low concentrations of radioactivity contamination to be "below regulatory concern" (BRC) and hence deregulated (55 Federal Register 27522; July 3, 1990). However, Congress intervened to set aside the BRC policy in the Energy Policy Act of 1992 after the USNRC's own suspension of the policy (56 Federal Register 36068; July 30, 1991). Circumstances considered for clearance (unrestricted release) include materials in which radioactive contamination is so low that clearance is warranted. In contrast to the release of a material from regulatory control, *exemption* from control may be considered in some circumstances, for example, when a small amount of radioactive material is added to a product deliberately to serve some justified purpose.

To account for different possible exposures, the exposure limit set for clearance (i.e., unrestricted release) or exemption of a material would have to be a small fraction of the 100 mrem/yr total limit. The revised Part 20 did not include specific standards for exemption; for case-by-case review, it is identical to the previous version. The 1991 version of Part 20 contains no regulatory statement defining a floor for regulated radionuclide content, other than the reference (noted above) to the NCRP recommendation on negligible individual risk of 10 µSv/yr (1 mrem/yr).

RADIATION PROTECTION STANDARDS DEVELOPED BY ORGANIZATIONS OTHER THAN THE USNRC

Organizations in Europe have developed basic radiation safety standards. Beginning in 1982, the International Atomic Energy Agency (IAEA) published a number of recommendations. Appendix C reviews IAEA Safety Series 89 along with safety standards developed by a number of other agencies. All of these standards recommend an individual dose on the order of 10 µSv/yr (1 mrem/yr) as the basis for clearance of materials from regulatory control.

Dose Comparisons

Standards for releasing SRSM are often based on a small percentage of the dose that a member of the U.S. population receives from what is termed background radiation (see definitions in Box 1-1). Table 1-1 lists the average annual dose to an individual in the United States from both natural and anthropogenic sources of ionizing radiation.

The values in Table 1-1 are averages, and the levels of background radiation are not uniform for individuals in different locations and having different lifestyles (see Table 1-2). A person living at higher altitudes receives more cosmic radiation than someone living near sea level. (For example, a person living in Denver, Colorado, receives 200 µSv/yr [20 mrem/yr] more than a person living on the Atlantic Seaboard, but when all natural sources are included the difference is 600 µSv/yr [60 mrem/yr] [NCRP, 1993].) A person living in a brick house receives an annual dose that is 70 µSv (7 mrem) higher than the dose for a person living in a frame house. An individual flying across the country receives a dose of about 25 µSv (2.5 mrem) per flight.

THE U.S. AND GLOBAL CONTEXTS OF RADIOACTIVE WASTE GENERATION

The ionizing radiation from radioactive materials has been used for more than a century. X rays and radium were soon used in the radiation treatment of cancer. Nuclear medicine followed, when radioactive tracers became available in 1931, after the development of the cyclotron. Nuclear weapons were developed during World War II, and the industrial processes involved also produced large quantities of radionuclides with long half-lives. Nuclear power plants to generate

TABLE 1-1 Average Annual Amounts of Ionizing Radiation to Which Individuals in the United States Are Exposed

Source	Dose mSv/yr	mrem/yr	Percent of Total Dose
Natural			
Radon	2.0	200	55
Cosmic	0.27	27	8
Terrestrial	0.28	28	8
Internal	0.39	39	11
Total Natural	3.0	300	82
Anthropogenic			
Medical[a]			
X-ray diagnosis	0.39	39	11
Nuclear medicine	0.14	14	4
Consumer products	0.10	10	3
Occupational	<0.01	<1.0	<0.03
Nuclear fuel cycle	<0.01	<1.0	<0.03
Nuclear fallout	<0.01	<1.0	<0.03
Miscellaneous	<0.01	<1.0	<0.03
Total anthropogenic	0.63	63	18
Total natural and anthropogenic	3.6	360	100

SOURCE: NCRP (1987a).
[a]UNSCEAR (2000) reports 1.2 mSv as the average medical dose for health care level I countries.

TABLE 1-2 Common Sources of Radiation to Which the Public Is Exposed

Source	Dose Equivalent (μSv) (mrem)
One-way, transcontinental or trans-atlantic airplane flight at mid-latitudes	25 (2.5)
Gas mantles (containing thorium), 1 year's typical use	2 (0.2)
Additional annual dose received from residence in a brick house, versus a wooden frame house	70 (7)
Annual dose from nuclear power plant to maximally exposed person (airborne effluents)	
Pressurized water reactor	6 (0.6)
Boiling water reactor	1 (0.1)
Annual dose received from natural levels of potassium-40 in the body	180 (18)
Additional annual dose from cosmic rays received in Santa Fe, New Mexico, versus sea level	450 (45)
Additional annual dose from natural background received in Denver, Colorado, versus Atlantic Seaboard due to all natural sources (cosmic rays, terrestrial deposits of radionuclides, etc.)	600 (60)

SOURCES: NCRP (1987a, 1987b, 1993); NRC (1999).

electricity soon followed, and over a period of about 30 years the power industry added nuclear capacity to coal, natural gas, and other sources of energy used to generate electricity. In the United States, 103 nuclear power reactor units now produce about 20 percent of the nation's electricity.

Soon after the United States developed nuclear weapons and nuclear power reactors, the developed nations in Europe and Asia followed with their own nuclear development programs. Nuclear power reactors are now used widely to generate electricity in many countries. (In France, approximately 80 percent of the electric power requirements are generated with nuclear fuel.) With the global spread of nuclear weapons and nuclear power, large quantities of radioactive materials have been generated in both developed and developing countries, and the global distribution of radioactive material raises important considerations. With global trade, at least trace amounts of radioactive materials will certainly be shipped across many borders. Detailed discussion of the international aspects of clearance regulations can be found in Chapter 7.

Radioactive waste is generated by many different industries and is regulated within the United States by several federal agencies, with the general exception of naturally occurring radioactive material (NORM) and naturally occurring and accelerator-produced radioactive material (NARM).[10] The larger sources (generators) of regulated radioactive materials are listed below:

1. Licensees of the USNRC and agreement states,
2. U.S. Department of Energy (DOE),
3. U.S. Department of Defense (DoD), and
4. Domestic nonnuclear industries[11] that nevertheless accumulate process wastes with significant radioactive material content.

The control and release practices of each of these generators (or generator categories) are discussed in subsequent subsections. These practices are important to considerations of alternative disposition approaches.

The USNRC System

The USNRC regulates radioactive materials through licenses. Among the licensees are many thousands of small users of sealed sources,[12] about a thousand

---

[10]DOE guidance applies to the management of NORM at its own facilities, but the regulation of NORM and NARM is otherwise performed only by states under applicable state law.

[11]By "nonnuclear industry," the committee means an industry whose processes are neither based upon nor designed to make use of radionuclide decay or fission reactions. Thus, an industry in which radioactive material may accumulate as an unsought concomitant of the industrial processes being used, such as petroleum drilling or phosphate mining, is a nonnuclear industry.

[12]Sealed sources are byproduct material encased in a capsule to prevent leakage. They typically contain a concentrated form of one radionuclide (e.g., ^{137}Cs).

hospitals, 104 licensed nuclear power reactor units (of which 103 are operating), 36 operating nonpower reactor units, 49 fuel cycle facilities, and 5,288 materials licensees. Agreement states have issued an additional 15,512 materials licenses (SCA, 2001). Generation of SRSM is generally not an issue for licensees using sealed sources—provided the sources are maintained in a safe condition and location.[13] For all licensees, the primary disposal issue is access to disposal options at reasonable cost. For USNRC licensees, most of the SRSM inventory (metals, concrete, soils, equipment, etc.) that may undergo clearance is associated with operating or decommissioning the 104 nuclear power reactor units at 65 sites, which are distributed across the country, with 32 states having one or more units.

In principle, the schedule for decontaminating and decommissioning a nuclear power reactor unit is established by the terms of its operating license. However, because the economics of nuclear power production in the United States have changed dramatically in recent years for a variety of reasons, the trend among licensees is to apply for extensions to their licenses. Because the development of these power plants was closely regulated from the industry's inception, the location, types, and amounts of contamination associated with these plants are known.

Procedures for decommissioning reactors have already been established, based on three options: decontamination, safe storage, or entombment. Some of the alternative approaches to the disposition of SRSM could facilitate decommissioning by markedly reducing costs.

The DOE System

Inventories of contaminated metal scrap have been identified at 13 DOE sites. Although not licensed by the USNRC, DOE manages and disposes of a significant portion of the nuclear material within the United States and is discussed here to show the broader context for the handling and disposition of such material. Because most DOE sites were involved in producing enriched uranium and plutonium, the radioactive materials contain long-lived radionuclides, including actinides such as neptunium and americium. DOE operated 14 plutonium production reactors at the Hanford Site and the Savannah River Site, producing about 100 tons of ^{239}Pu, which has a half-life of 24,390 years. Chemical separa-

---

[13] Although contamination from maintained sealed sources is not an issue, some sealed sources are lost. If these lost sources, known as orphan sources, enter the scrap metal stream, they pose a serious problem for the steel industry. Orphan sources in the scrap stream are difficult to detect. If by accident they are melted into the production stream, major sections of a steel mill can be contaminated, causing tens of millions of dollars of damage.

tion processes for the recovery of plutonium and uranium generated more than 100 million gallons of radioactive wastes, which are currently stored at several DOE sites (SCA, 2001).

The DOE sites are large—often measured in hundreds of square miles. For example, the Hanford Site is about 560 square miles and the Savannah River Site is approximately 310 square miles. Production facilities at these large sites occupy only a small fraction of the total site area. Because many of the sites are well removed from populated areas, long-term on-site storage or burial has been one option employed for handling wastes. In addition, the Savannah River Site and the Nevada Test Site[14] are currently used for disposal of DOE-generated LLRW.

The facilities at most DOE sites are large relative to most industrial plants. For example, the K-25 gaseous diffusion plant, built in 1943 at Oak Ridge, Tennessee, is a three-level building that occupies 44 acres. In many instances, the DOE facilities are no longer functioning but still contain significant amounts of SRSM. Also, some of the equipment used to produce weapons-grade materials is classified and must be deconfigured at secure sites before disposal.

Production activities at many of the DOE sites began in 1943, when the dangers of ionizing radiation were less well understood or perhaps not of greatest concern. In a climate of wartime urgency, creating an entirely new and huge production complex and running it at full capacity were the critical concerns. Materials were disposed or stored on-site, with limited attention to the safeguards now taken for granted. Today, cleaning up discarded radioactive materials from the 1940s and 1950s at many DOE sites poses major problems for the contractors involved. The projected costs are enormous. Due to the complex history of defense-related operations at DOE facilities, material and waste management practices varied widely over the past half-century. This history often complicates the application of criteria for the release of solid materials during decommissioning of DOE facilities.

The DoD System

The DoD system includes both USNRC-licensed operations, covering a spectrum of operations similar to those found in the civilian world, and assets related to the nuclear Navy. The DoD facilities licensed by the USNRC include hospitals, laboratories, proving grounds, some nuclear reactors, weapons facilities, and missile launch sites. The USNRC does not license the nuclear Navy's assets, which include naval nuclear reactors and associated propulsion units. When nuclear-powered vessels are decommissioned, the reactor compartments are cut from the hull, sealed, and shipped to the DOE Hanford Site for burial. The ship

---

[14]This site, formerly used for nuclear weapons tests, is the largest in the DOE complex and occupies about 1,350 square miles in a remote area about 65 miles northwest of Las Vegas.

hulls are scrapped. The guidelines followed for clearing materials for reuse or recycle are classified. As of April 1999 the U.S. Navy had shipped 79 reactor compartment packages (representing 77 submarines and 1 cruiser) to the Hanford Site for disposal. There are about 2,800 tons of various types of recyclable metals in a submarine and 6,000 tons in a cruiser (SCA, 2001). Thus, more than 220,000 tons of steel, aluminum, copper, lead, and other metals have been recycled or reused from the Navy's decommissioning efforts.

About 115,000 cubic feet of LLRW is generated annually from DoD facilities. Most of this waste is from cleanup efforts rather than operations. As a group, the USNRC-licensed facilities of DoD appear to raise no unique inventory issues.

Non-USNRC-Licensed Industries

Among the U.S. industries that generate radioactive solids are several that can be described as nonnuclear because the processes employed do not intentionally use nuclear decay or nuclear fission reactions. Among these industries are petroleum production and refining, phosphate and phosphate fertilizer production, coal-fired power plants, and mining. The wastes generated contain NORM or technically enhanced NORM (TENORM). The USNRC estimates that more than 2 million metric tons of TENORM are generated annually (USNRC, 2001a). Much of this material contains significant concentrations of uranium, thorium, and radium radionuclides, all of which have long half-lives.

There are no federal statutes that specifically establish regulatory control of TENORM, although some waste streams fall under the jurisdiction of the EPA. Control of TENORM has been left to the states, and some agreement states regulate TENORM under their general rules governing the possession of radioactive materials. In many states with agreement state authority, the regulation of NORM, TENORM, and NARM comes under the same program used to regulate radioactive materials controlled under the Atomic Energy Act (AEA).

About 75 Superfund sites are contaminated with radioactive wastes[15] (Wolbarst et al., 1999). Many of these are DoD and DOE sites, but more than 20 were created by commercial industrial waste disposal.

STATUS OF THE CURRENT USNRC PROCESS FOR CLEARING SOLID MATERIALS

The USNRC has statutory responsibility for the protection of public health and safety related to the use of source material, byproduct material, and special

---

[15]"Superfund" is the commonly used term for the Comprehensive Environmental Response, Compensation, and Liability Act.

nuclear material, as defined by the AEA.[16] The USNRC's regulations in fulfillment of these goals include those on protection against radiation (10 CFR Part 20 et seq.), licensing of byproduct material (10 CFR Part 30 et seq.), licensing of source material (10 CFR Part 40), licensing of production and utilization facilities (i.e., nuclear reactors; 10 CFR Part 50 et seq.), licensing of special nuclear material (10 CFR Part 70 et seq.), and so forth.

As noted, the regulations on protection against radiation, 10 CFR Part 20, do not set predetermined levels on amounts or quantities of radionuclides in solid materials below which these materials can be released from further regulatory control. Solid materials potentially available for release from regulatory control include metals, building concrete, on-site soils, equipment, and furniture used in routine operation of licensed nuclear facilities. Most of this material will have no radioactive contamination, but some of it may have surface or volume contamination. Licensees continue to request permission from the USNRC and agreement states to release such solid materials when they are no longer useful or when the licensed facility is decommissioned, pursuant to Section 2002 of 10 CFR Part 20. In addition, as noted, Regulatory Guide 1.86 (AEC, 1974) contains limits applicable to surface contamination that are incorporated into license conditions and allow clearance of SRSM.

The USNRC allows licensees to release solid material according to preestablished criteria. For reactors, if surveys for surface residual radioactivity performed by the licensee on equipment or material indicate the presence of radioactivity above natural background levels, then release is not permissible.[17] If no such surface activity is detected, then the solid material in question need not be treated as radioactive material. This approach sometimes leads to subsequent problems, when detectors of greater sensitivity than were used in the initial survey detect radioactivity above the natural background threshold in previously released material (USNRC, 2001b).

For surface-contaminated SRSM possessed by a materials licensee, the USNRC usually authorizes its release through specific license conditions or technical specifications (USNRC, 2001b). In the case of volume-contaminated SRSM held by reactor and materials licensees, the USNRC has not provided guidance similar to that found in Regulatory Guide 1.86 for surface contamination. These situations are decided instead on an individual basis pursuant to Section 2002 of 10 CFR Part 20, typically by evaluating the doses likely to be associated with the proposed disposition of the material. The case-by-case approach has some distinct advantages and disadvantages, as discussed in Chapter 2 and Chapter 9.

The Commission directed the USNRC staff in June 1998 to consider a rulemaking for establishing a dose-based standard for release of SRSM (USNRC,

---

[16]Chapter 2 discusses the AEA definitions of these materials.

[17]Reactor licensees can apply to UNSRC for approval for clearance of solid materials with small but detectable levels of radioactivity pursuant to Section 2002 of 10 CFR Part 20.

INTRODUCTION 27

1998a). The intent was to provide for consistent disposition of SRSM while protecting public health and safety. The USNRC staff was also directed to ensure that opportunities would be provided under the proposed standard for enhanced public participation. The USNRC subsequently published an issues paper outlining possible courses of action were it to proceed with a rulemaking (64 Federal Register 35090-35100; June 30, 1999). As a first option, according to the issues paper, the USNRC could restrict the release of SRSM only for certain authorized uses or disposition options, in which the potential exposure to the public would be small (conditional clearance). For example, restricting the options to disposal of the SRSM in Resource Conservation and Recovery Act (RCRA) Subtitle D landfills[18] is a conditional clearance that would significantly reduce the number of exposure pathways, relative to a situation in which the material is recycled into consumer products. As a second option, the USNRC could permit the release of solid materials for unrestricted use if the potential for exposure to the public from projected uses were less than a specified dose level (clearance). Unrestricted use might include recycle or reuse of SRSM in consumer or industrial products or any other use. As a third option, the USNRC could prohibit both unrestricted and restricted release of SRSM from a licensed facility. Instead, it could require that such material go to an LLRW facility. For each of these alternatives, the impacts on public health and the environment, as well as on cost-benefit factors, should be considered. Consideration of the means of implementing each alternative and its practicality would also be important if a rulemaking is undertaken.

The issues paper notes that consideration of rulemaking alternatives for solid material release would cause the USNRC to examine the existing policies of international bodies, other federal agencies, state governments, and other standard-setting bodies. The IAEA and the Commission of European Communities have made significant efforts to set standards for the release of SRSM. These bodies have adopted sets of standards based on an annual dose of 10 µSv/yr (1 mrem/yr), which is broadly accepted by the radiation protection community as a de minimis dose.[19] Consistency among standards is an important concern because of the potential import or export of released materials between the United States and other countries.

The issues paper further notes the importance of coordination with other federal agencies, such as the EPA. In regulating its licensees, the USNRC imple-

---

[18]RCRA defines under separate subtitles the land disposal requirements for categories of waste at different levels of potential health or environmental hazard. Subtitle D covers the lowest level of potential hazard—wastes equivalent to general municipal waste. Landfills meeting these requirements are called Subtitle D landfills. Similarly, landfills suitable for most common hazardous materials generally used in or produced by industry are regulated under Subtitle C and are called Subtitle C landfills.

[19]A de minimis dose is one at or below which statutory or regulatory controls would not apply. The legal term "de minimis" is shorthand for *de minimis non curat lex*, which is Latin for the common law doctrine stating, in free translation, that "the law does not concern itself with trifles."

ments the environmental standards set by the EPA. In the absence of EPA standards in areas such as the release of SRSM, the USNRC has the authority to set standards. If proposed USNRC actions are not closely coordinated with the EPA, problems could develop if the EPA later adopted conflicting standards. A majority of the states have entered into agreements with the USNRC to assume regulatory authority over small quantities of byproducts, sources, and nuclear material. Other standard-setting bodies such as the NCRP could play important roles in setting dose standards for release of solid materials. The NCRP, a nonprofit corporation chartered by the U.S. Congress, makes recommendations regarding acceptable levels of radiation exposure to the general public, including levels considered to present a de minimis health risk.

THE STUDY TASK AND APPROACH

The USNRC is considering whether to establish a new regulation that would set specific limits for the release of solid materials with low levels of radioactivity (64 Federal Register 35090-35100; June 30, 1999). The primary reason for a new regulation would be to provide consistency in USNRC's regulatory framework for releases of solid materials, including materials with volume contamination. Standards for the release of radioactively contaminated gaseous and liquid materials have already been established.

The USNRC has sought public input in contemplation of such a rulemaking. Two-day meetings were held in Chicago, San Francisco, Atlanta, and Rockville, Maryland, in late 1999. Hundreds of written and electronic comments from the public at large were received. Following the public meetings, the USNRC contracted with the National Academy of Sciences to study several critical issues related to the release of solid materials with low levels of radioactive contamination. The statement of work, which appears in excerpted form below, outlines five tasks, to be performed by a committee appointed in accordance with the procedures of the National Research Council (see Appendix C for the complete statement of work):

1. As part of its data gathering and understanding the technical basis for the Nuclear Regulatory Commission's (USNRC's) analyses of various alternatives for managing solid materials from USNRC-licensed facilities, the committee shall review the technical bases and policies and precedents derived therefrom set by the USNRC and other Federal agencies, by States, other nations and international agencies, and other standard setting bodies.
2. The committee will review public comments and reactions received so far on current and former USNRC proposals to develop alternatives for control of solid materials. The committee will explicitly consider how to address public perception of risks associated with the direct reuse, re-

cycle, or disposal of solid materials released from USNRC-licensed facilities. The committee should provide recommendations for USNRC consideration on how comments and concerns of stakeholders can be integrated into an acceptable approach for proceeding to address the release of solid materials.
3. The committee shall determine whether there are sufficient technical bases to establish criteria for controlling the release of slightly contaminated solid materials. This effort should include an evaluation of methods to identify the critical groups, exposure pathway(s), assessment of individual and collective dose, exposure scenarios, and the validation and verification of exposure criteria for regulatory purposes (i.e., decision making and compliance). As part of this determination, the committee should judge whether there is adequate, affordable measurement technology for USNRC-licensees to verify and demonstrate compliance with a release criteria. What, if any, additional analyses or technical bases are needed before release criteria can be established?
4. Based on its evaluation and its review, the committee shall recommend whether USNRC (1) continue the current system of case-by-case decisions on control of material using existing, revised, or new (to address volumetrically contaminated materials) regulatory guidance, (2) establish a national standard by rulemaking, to establish generic criteria for controlling the release of solid materials, or (3) consider another alternative approach(es).

If the committee recommends continuation of the current system of case-by-case decisions, the committee shall provide recommendations on if and how the current system of authorizing the release of solid materials should be revised.

If the committee recommends that USNRC promulgate a national standard for the release of solid material, the committee shall: (1) recommend an approach, (2) set the basis for release criteria (e.g., dose, activity, or detectability-based), and (3) suggest a basis for establishing a numerical limit(s) with regard to the release criteria or, if it deems appropriate, propose a numerical limit.
5. The committee shall make recommendations on how the USNRC might consider international clearance (i.e., solid material release) standards in its implementation of the recommended technical approach.

Limitations of the Study

In response to the USNRC request, the National Research Council established the Committee on Alternatives for Controlling the Release of Solid Materials from Nuclear Regulatory Commission-Licensed Facilities (hereafter, the

TABLE 1-3 Risk Assessment Based on a Linear, No-Threshold Model with a Probability of Developing a Fatal Cancer of 5×10^{-2} /Sv (5×10^{-4}/rem)

Incremental Dose	Hypothetical Incremental Lifetime Risk	Hypothetical Lifetime Risk (If dose received each year for 70 years)
1.0 mSv (100 mrem)	5×10^{-5}	3.5×10^{-3}
0.1 mSv (10 mrem)	5×10^{-6}	3.5×10^{-4}
0.01 mSv (1.0 mrem)	5×10^{-7}	3.5×10^{-5}
0.001 mSv (0.1 mrem)	5×10^{-8}	3.5×10^{-6}

"committee"). In completing the five tasks listed above, the committee has worked under several limitations and constraints that are worth noting at the outset. First, for determination of the risk assessments on the health effects of incremental doses, the committee has relied on assessments by UNSCEAR (1988), the National Research Council Committee on the Biological Effects of Ionizing Radiation (NRC, 1990) and the NCRP (1993). These assessments found that a lifetime risk[20] of developing a fatal cancer from low dose or low dose rate irradiation is estimated to be 5×10^{-2}/Sv (5×10^{-4}/rem) for an individual in the general population. Table 1-3 shows the risk estimates developed by NCRP (1993) by applying the linear, no-threshold hypothesis to various incremental annual doses.

Second, the committee did not independently explore the relative validity of various biological risk assessments associated with radiation dose. Such assessments for low doses are controversial. They are subject to the assumptions made according to the model employed. Independent evaluation of the validity of the various risk assessments was beyond the scope of the task before the committee.

A third limitation was the exclusion of soils from major consideration. The amount of soil involved in decommissioning the nuclear power plants is generally small relative to the quantities of concrete and metals as shown in Chapter 3 (Table 3-6). On the other hand, the amount of contaminated soil at DOE facilities can be significant.

Study Process

The committee organized three information-gathering meetings, at which speakers were invited to make presentations before the committee on a range of technical issues. Several stakeholder groups presented their views to the commit-

---

[20]Lifetime risk is the likelihood of an adverse health effect occurring (fatal cancer, in this instance) at *any* time in the future due to exposure to radiation.

tee. Views from industries affected by proposed clearance of SRSM were also presented. Meetings in which information was presented to the committee were open to the public, and when time permitted, either the speakers or members of the committee addressed questions from the audience. Speakers were encouraged to provide written statements or to provide the audience with copies of their visual aids. Appendix B gives a detailed account of the speakers who provided information to the committee at these meetings.

Certain members of the study committee visited two waste brokers, ATG in Richland, Washington, and Duratek, Inc., in Oak Ridge, Tennessee. The members observed and studied the methods currently used to release solid materials with low concentrations of radioactive contamination from regulatory control.

Report Content

The regulatory framework for controlling the release of solid materials with radioactive contamination is described in Chapter 2, which is organized into three main sections. The first deals with the technical assumptions underlying radiation standards and includes a review of the important concepts employed in establishing radiation standards. The second section discusses the historical evolution of regulatory practices and controls in greater technical detail than the introductory account in this chapter. The third section provides a comparative assessment of existing regulatory regimes in the United States.

Chapter 3 discusses the inventory of radioactively contaminated solid materials from USNRC licensees, DOE, DoD, and various industrial sources. The first section of the chapter deals with waste streams from nuclear reactors. The second section presents a much broader view of the accumulated inventory, including licensed fuel cycle and non-fuel cycle facilities, DOE, DoD, EPA Superfund sites, NORM, and TENORM.

Chapter 4 defines major alternatives for the disposition of solid materials with low concentrations of radioactivity. A decision diagram with decision points and disposition pathways is described. Estimated costs for various disposition alternatives are discussed because disposal costs are markedly affected by the disposal options available to a licensee—for example, which disposal sites can be used by a licensee for different categories of solid materials.

Chapter 5 reviews the technical basis for developing dose-based standards. Implementing a dose-based standard requires a conversion from a concentration of radioactivity in a solid matrix, as measured before release, to estimated doses resulting from exposure of an individual in a critical group to that material. The critical pathways and the assumptions made in performing these conversions are discussed, as are the uncertainties in determining the factors for converting between measurable radioactivity levels and a dose standard.

Chapter 6 discusses the difficulties in quantitatively determining the identity and activity of the radionuclides present in SRSM. It reviews the capability and costs of instrumentation and measurement procedures to conduct the determination at various proposed screening levels. Also discussed are current measurement practices of waste brokers and approaches to develop an appropriate sampling program.

Chapter 7 reviews the efforts to develop international clearance standards. The final section of the chapter summarizes the status of several countries in establishing clearance standards for the release of SRSM.

Chapter 8 reviews stakeholder concerns and issues regarding past and recent efforts of the USNRC to establish a clearance standard for SRSM. The chapter emphasizes the importance of effective risk communication and establishing trust in building stakeholder acceptance. Consensus-building processes to involve stakeholders are presented.

Chapter 9 presents the committee's version of a decision framework for considering alternatives for controlling the release of solid materials with radioactive contamination. First, the problems with the current USNRC approach are described. Then a systematic decision framework for considering the alternatives for release of radioactive material is presented. The chapter also addresses issues of public perception. A section on process considerations provides options for obtaining enhanced participation from the public and possibly proceeding to a rulemaking.

Chapter 10 contains key findings from the report that serve as a foundation for the committee's recommendations. The committee's recommendations are presented as well.

2

The Regulatory Framework

MECHANICS OF EXISTING AND FORMER STANDARDS GOVERNING RELEASES OF RADIOACTIVELY CONTAMINATED MATERIAL

The technical assumptions underlying existing and former radiation standards are integral to the standards themselves and thus critical to evaluating them. In this section, the study committee reviews several of the most important concepts used in establishing radiation standards, including dose-based versus activity-based standards, the role of calculated simulations in assessing risks, the importance of defining critical groups, and important uncertainties in assessing risks (see Box 2-1 for description of different types of radiation standards).

The general trend in environmental regulation is toward risk-based standards, which typically focus on the estimated increased lifetime risk of cancer posed by the regulated material (NRC, 1994). Certain statutes, however (e.g., sections of the Clean Air Act), continue to use technology-based standards. That is, they prescribe the use of a particular control technology rather than establishing an acceptable exposure level. Calculating the health risks associated with a radioactively contaminated object involves a two-step process. First, the dose must be calculated, which entails constructing a range of scenarios to represent the range of potential doses to individuals. Second, for each estimated dose, the attendant health risk or harm must be estimated. As discussed below, both steps necessarily introduce uncertainties and typically use simplifying assumptions.

The virtues of a risk-based approach are that it establishes standards close to the level of public health concern, ensures that contaminant levels are controlled

> **BOX 2-1**
> **Different Types of Radiation Standards**
>
> Radiation standards are set to protect the public from harmful exposure to radiation from the direct radiation, releases or residues of nuclear materials. Different concepts are used in establishing the radiation standards; the different types are named after the concept on which they are based:
>
> - *Technology-Based Standard.* A technology-based standard requires application of best available technology to reduce exposure to acceptable levels, levels that are the lowest reasonably achievable.
> - *Risk-Based Standard.* A risk-based standard requires measurement against a designated level of exposure that defines acceptable risk. A simple example of a risk-based standard is one that sets a limit for chronic radiation exposure at a level that is associated with an acceptable range of probability for the lifetime risk of cancer. A more complex example of a risk-based standard is found in 10 CFR Part 63, the regulation for safe disposal of high-level radioactive waste. This requires that the results of the probabilistic performance assessment for Yucca Mountain be compared directly to the compliance standards set in Part 63.
> - *Risk-Informed Standard.* A risk-informed standard is one in which specific bases for acceptance are set in the standard and a separate risk assessment is used to examine whether the standards were prudently chosen. Acceptance is not based on direct compliance with risk terms. Perhaps the broadest example of risk-informed regulation is the USNRC's complex code of standards for licensing nuclear power plants, where specific probabilistic risk assessments (PRAs) are used in supplement to examine the residual risk for plants licensed against these standards. The results of the specific PRAs may be compared to published safety goals, but the plants are not directly licensed to comply with these safety goals. The reactors must comply with the panoply of specific requirements for licensing; the PRA provides a supplemental analysis to estimate whether the reactor achieves the safety goal.
> - *Dose-Based Standard.* A dose-based standard is one that sets the maximum radiation exposure from a source, for example, from released slightly radioactive solid material, that might be suffered by the most exposed group in the public.
> - *Activity-Based Standard.* An activity-based standard is one that sets limits on the radioactive content of a source, for example, from released SRSM, where the limits are derived from acceptable exposure rates.

to achieve acceptable levels of public health protection, and promotes consistency among different regulations. Risk-based standards are meant to be responsive to public policy decisions on widely acceptable levels of risk and are presumed to be rationally based on carefully conducted estimates of dose and risk. The unavoidable uncertainties in risk-based standards are therefore more than offset by their capacity to incorporate policy determinations into a rigorous, scientifically based framework. However, an important challenge is to ensure that the methods used, including their simplifying assumptions and inherent constraints, are sufficiently transparent to both technical peers and the concerned public.

As noted earlier, two types of standards exist in the area of radiation safety for slightly radioactive solid material (SRSM). One type is based on the level of radiation exposure, or dose. The other is based on the level of radioactivity of the material in question and is therefore often called an activity-based standard. Superficially, a radioactivity-based standard appears to be the more direct of the two approaches because it prescribes a maximum level of radiation that may be emitted by an object that is to be used or disposed in a specified manner. A radioactivity-based standard does not appear to require the complex process of assessing how individuals might be exposed to the object's radioactivity and what the resulting doses are likely to be. Technical analyses such as draft NUREG-1640 (USNRC, 1998b) derive radioactivity-based limits for selected disposition cases that are based on risk or dose limits (see Chapter 5). However as discussed further below, whether there is in fact a significant difference in complexity between these two types of standards depends on whether the governing regulation is based on technology (i.e., a control or measurement limitation) or on limiting exposure, hence risk.

Technology-Based Regulations

Regulatory standards may be based on the limitations of existing control or measurement technologies. The U.S. Nuclear Regulatory Commission's (USNRC's) existing guidance document concerning release of solid materials with surface contamination from regulatory control, developed in the 1970s, is based on the decontamination survey practices that were in use at that time (see Box 2-2). Some environmental laws, such as specific provisions in the Clean Air Act, base regulations on the "best available control technologies." In this approach to regulation, the focus is not on risk, which is difficult to estimate and even harder to defend, but on promoting the use of the most advanced technologies and fostering their further development.

Regulations that require the use of best available control technology obviate the need for dose estimates. In some instances, specifying activity limits is not necessary. The salient issue is maximizing the use of the most effective control technologies. To achieve this, a regulation could prescribe limits on radioactivity levels (e.g., annual emissions limits on radionuclides) or require that specified instruments or methods, and defined limits, be employed when radioactively contaminated materials are monitored. To a large degree, the existing guidance embodies this latter principle, relying on extensive guidance for procedures and practices (AEC, 1974; USNRC, 1981).

Technology-based regulation has the advantage of being relatively simple to implement. It avoids the complexities of determining the myriad ways in which

> **BOX 2-2**
> **Regulatory Guide 1.86: Guidance for Unrestricted Release**
>
> Regulatory Guide 1.86, *Termination of Operating Licenses for Nuclear Reactors*, was published in June 1974. In addition to guidance on reactor license amendments, it included an important section, "Decontamination for Release for Unrestricted Use," which established the guidelines for reactor decommissioning and the clearance of solid materials. This section included a table that codified established standards at many sites for adequate decontamination of surfaces. The key language for the present purposes is contained in Section 4:
>
>> After the decontamination is satisfactorily accomplished and the site inspected by the Commission, the Commission may authorize the license to be terminated and the facility abandoned or released for unrestricted use. The licensee should perform the decontamination using the following guidelines [paraphrased]:
>>
>> - A reasonable effort should be made to eliminate residual contamination.
>> - No covering should be applied to radioactive surfaces.
>> - The radioactivity of the interior surfaces should be determined.
>> - The USNRC may authorize controlled release to another licensee based on detailed health and safety analysis of premises, equipment, and scrap.
>> - Prior to release for unrestricted use, the licensee should report the results of a comprehensive radiation survey.
>
> SOURCE: AEC (1974).

people might be exposed to radiation from radioactively contaminated materials. A major disadvantage, however, is that if the approach were applied in total ignorance of the potential harms, it could result in either serious underregulation and thus increased risk to the public or overregulation and hence increased costs to the regulated industries. Thus, when developing technology-based regulations, regulatory agencies are well advised to conduct at least brief analyses of the risk reduction and cost-benefit achieved by the specific technologies that might be implemented.

Risk-Based Regulations

In practice, many standards are a hybrid of dose-based and activity-based approaches. For example, any risk-based standard, whether its allowed maximum levels are expressed as doses or radioactivity levels, entails that the ultimate dose to a certain class of individuals, termed the "critical group," be assessed. To bound the analysis in the assessment requires fairly elaborate simulations and numerous technical judgments. The inherent uncertainty associated with these simulations and judgments varies with the quality of data and the range of potential exposure scenarios that must be considered.

Constructing Critical Groups and Exposure Scenarios

Often, relatively clear bounding hypotheses for the analysis can be identified by using conservative assumptions about possible routes of exposure. For example, in developing its analyses for risk-based standards, the Environmental Protection Agency (EPA) has sought to identify plausible examples from which significant exposures could arise. It uses these exposure scenarios to construct the critical groups for the analysis. The doses to these critical groups, estimated by simulating the exposure scenario, dictate the level of radioactivity that is permitted in materials subject to regulation.

As an illustration of how critical groups are used, one critical group considered by the EPA is represented by an operator of an industrial lathe made with radioactively contaminated cast iron. This is a relatively high-dose scenario because of the time spent next to the radioactive object, as well as its size and proximity. The larger the object, given the same concentration of radioactive material, and the longer the time in proximity, the higher is the exposure (EPA, 1997a).

In most cases, as in the above example, the doses to the critical groups are constructed to provide the upper bound on what is permissible under the regulation. The method assumes that most of the public will be exposed to far lower levels of radiation than would members of the critical groups.

An important question that is frequently raised about such simulations is whether exposure from a number of different sources could lead to much higher levels of risk. Returning to the example of the lathe operator, multiple exposures would occur if this individual were exposed not only to radiation from the industrial lathe but also to radiation from cast iron cooking utensils and large home appliances. In theory, these multiple routes of exposure could raise the individual's exposure above the level that the applicable regulation is attempting to ensure is not exceeded.

Regulators work to account for the potential that multiple exposures will occur by using information on the volume of materials at issue, the materials' potential uses, the relative importance of different routes of exposure, and the circumstances under which the materials are used (EPA, 1997a). Using this information and a conservative set of assumptions, regulators attempt to assess the likelihood and importance of multiple exposures. For completeness, it is important to take into account the potential for such multiple exposures, even where the levels of contamination in the materials are very low, since multiple exposures can result in a higher dose to an individual than originally analyzed. Allowance for multiple exposures may be in the form of choosing a level for a standard that reflects the likelihood of multiple exposure. Thus, the standard for release of a site may be a relatively large fraction of the public exposure safety limit, while the standard for release of material into commerce would be a much smaller fraction, even a de minimis level.

Uncertainty and Sensitivity of Analytical Assumptions

The inherent complexity of dose assessment analyses requires that numerous simplifying assumptions be made. For example, assumptions must be made about the length of time a person spends next to a contaminated object and at what distance, as well as whether the contaminated material is mixed with clean materials before being fabricated into a consumer product. These assumptions and the variability in the quality of information available mean that the exposure simulations on which the analysis depends are subject to significant uncertainties. These uncertainties are typically difficult to quantify. If overly conservative assumptions are used in the analysis, the assessment will err on the side of caution. Conversely, if simplifying assumptions minimize or underestimate potential risks, the assessment will err toward inadequate control to protect health and safety. If uncertainty distributions or ranges for the input assumptions are available, analysts can perform studies, using methods such as Monte Carlo simulations, to obtain estimates of the uncertainties in the dose calculations or other predictions from the analysis (see Chapter 5 for further discussion).

In addition to uncertainty, the difference that any given assumption makes to the overall analysis can be quantified by using a *sensitivity analysis*. The sensitivity of the final dose estimate to a particular input assumption or factor is measured by varying the value assumed for that assumption without varying any other factors.

Although Monte Carlo simulations and sensitivity analyses can be complicated by variables that are strongly dependent, they provide an important means by which analysts can gain a qualitative sense of the reliability, or variability, of their estimates and an understanding of what factors are most important. Regulators therefore have at their disposal an array of analytical methods that can be used to assess whether their judgments are reasonable.

Critical Uncertainties

Although the analytical methods employed by regulators in establishing standards have become increasingly sophisticated, uncertainty and judgment are unavoidable in assessing potential risks and deciding how much extra conservatism to embed in the regulations. In the present context, there are several particularly important uncertainties, which the committee discusses at several points in this report. Among these uncertainties are the following:

- The risk that radionuclides will concentrate in certain solid materials that

are unconditionally released into commerce;
- The limits on existing radiation monitoring equipment and survey methods;
- The significance of multiple potential exposure pathways for cumulative exposure to the public; and
- The reliability of conservative, or bounding, hypotheses in designating critical groups.

The consequences of these uncertainties for assessing risks associated with radionuclides are particularly complex because many radionuclides are long-lived and because monitoring for low levels of radioactivity requires sophisticated instrumentation and rigorous methods. In making conservative estimates, regulators must carefully take these factors into account. As discussed in Chapter 5, analysts attempt to incorporate these factors into their calculations and assess their significance, at least qualitatively.

HISTORICAL EVOLUTION OF THE REGULATORY FRAMEWORK FOR CONTROLLING RADIOACTIVELY CONTAMINATED SOLID MATERIALS

Under the Atomic Energy Act of 1946 as amended in 1954 (AEA), the Atomic Energy Commission (AEC) and its successor agency, the USNRC, were granted the authority to regulate radioactive materials associated with nuclear fission. These materials are categorized in the AEA as *source materials* (i.e., uranium and thorium), *special nuclear materials* (e.g., plutonium), and *byproduct materials* (e.g., most radioactive material including common radioactive wastes) (42 U.S.C. §§ 2073, 2091, 2111).[1] Byproduct material includes any radioactive material (except special nuclear material) yielded in or made radioactive by the process of nuclear fission. This process includes both fission fragments (fission products) and activation products (42 U.S.C. § 2014(e)).[2] Notably, the AEA does cover naturally radioactive source materials, but does not cover naturally occurring radioactive material (NORM) (e.g., radon gas), technologically enhanced

---

[1] References to the United States Code (U.S.C.) are given parenthetically using the conventional format with the title number first (Title 42 in this reference), followed by the initials U.S.C. and the section numbers within the title.

[2] The Uranium Mill Tailings Radiation Control Act of 1978 (Public Law 95-604) added a second category of byproduct materials at section 11(e)(2) of the AEA, defining them as the "tailings" or waste produced by the extraction or concentration of uranium or thorium from any ore processed primarily for its source material (i.e., uranium or thorium) content. This and other terms have been paraphrased from their original sources, the Atomic Energy Act and 10 CFR Part 20. These sources should be consulted with regard to the precise legal meaning and effect of these terms.

NORM (TENORM), or materials made radioactive from particle accelerator experiments.

In establishing the AEC's regulatory authority, the AEA delineated appropriate regulatory procedures in substantial detail (42 U.S.C. §§ 2073, 2091, 2111). It did not prescribe specific technical requirements, deferring instead to the AEC, and later the USNRC, to develop and promulgate requirements for specific activities. Accordingly, all activities that were to be licensed by the AEC originally required the applicant to submit technical justifications for the proposed practice and to undergo a case-by-case review for authorization. Over time, specific requirements have been established for recurring or routine license applications.

Regulatory Practices and Controls

Title 10 (Energy) of the Code of Federal Regulations establishes licensing requirements for all practices using nuclear materials under the jurisdiction of the USNRC and agreement states. Examples include 10 CFR Part 40 for source material, 10 CFR Part 50, et seq., for facilities that produce or utilize special nuclear material, and a series of regulations beginning with 10 CFR Part 30 for byproduct material. These regulations codify licensing requirements in a generically applicable way to the extent possible.

The USNRC issues two basic types of licenses, specific and general. A *specific license* is required for practices involving significant quantities of nuclear material that warrant licensee control employing at least one radiation control professional. Commercial nuclear power plants, for example, are operated under a specific license issued by the USNRC. A *general license* may be issued if the quantity of nuclear material is significant but adequately protected through design and administrative controls (e.g., an industrial gauge that uses a strong radiation source). General licensees are not required to have radiation control professionals but are required to use a generally licensed device under the specified controls. The design and administrative controls are imposed through the specific licensee who makes and distributes a generally licensed device, as well as the end user of the device.

Certain radioactive materials may be deemed exempt from regulation if the amount of radioactive material involved is small enough or adequately protected by design. Examples include ionization smoke detectors and the small quantities and concentrations listed as exempt in 10 CFR §§ 30.70, 30.71.

Extensive regulations govern the disposal of radioactive wastes generated by or from licensed facilities. The regulations for high-level radioactive waste, 10 CFR Part 60 and Part 63, define high-level radioactive waste by its origin, not its radioactive content, and delineate detailed requirements for its licensed disposal. The disposal requirements for the three classes of low-level radioactive waste (LLRW) are contained in 10 CFR Part 61. Although these regulations impose upper bounds on the radioactive content for Class A, B, and C low-level waste,

THE REGULATORY FRAMEWORK   41

they do not specify a floor or threshold content of radioactivity below which material may be treated as nonradioactive waste. Accordingly, under existing regulations there is no generally applicable criterion for determining that the radioactive content in solid waste is de minimis.[3]

Formal USNRC regulations are augmented by a series of guidance documents, referred to as "Regulatory Guides," that establish preferred or acceptable methods for regulatory compliance purposes. Regulatory Guides are developed and proposed by committees of technical experts in a specific area, such as radiation monitoring or facility engineering requirements. If the USNRC endorses a proposed practice, it is formally published as a Regulatory Guide. Licensees who adopt a Regulatory Guide by incorporating it by reference in their license application are subject to inspection and enforcement of its requirements. A license applicant may choose instead to propose different practices for special reasons. However, doing so can lead to substantial delays in licensing decisions.

One such guidance document, Regulatory Guide 1.86, *Termination of Operating Licenses for Nuclear Reactors* (AEC, 1974), is of particular interest to the study committee's task (see Box 2-2). Issued in June 1974, this guide was released in the midst of the transition from the former AEC to the newly established USNRC. This was also the time when the first generation of demonstration power reactors was decommissioned. Unlike the typical document in this series, Regulatory Guide 1.86 was not developed by an expert committee; it was promulgated as a placeholder to enable reactor decommissioning to proceed. Thus, it enumerates licensing administrative requirements and different approaches to reactor decommissioning and specifies, in its fourth and final section, a systematic approach for license termination and release of equipment and the site.

Regulatory Guide 1.86 includes a table, Table I, of acceptable surface contamination levels. This AEC guidance for permitting clearance of radioactive materials dates back more than 25 years, to the initial preparation of Regulatory Guide 1.86. The Table I guidance had been in informal use for some time before 1974 and apparently was based on the detection limits of the instruments available at that time, not on an assessment of risk.[4] Table I contains guidance on clearance standards for surfaces such as floors, walls, structural materials, and equipment; it contains no standards for volume contamination. The table, which became the USNRC's de facto standard for clearance of solid materials with residual surface contamination, has been widely used for decades.

Selecting a clearance level requires that specific implementing protocols be developed. Office of Inspection and Enforcement (IE) Circular No. 81-07, *Con-*

---

[3]For two radionuclides only, in solid materials, and in one specific application, section 2005(a)(2) of 10 CFR Part 20 does contain release criteria. These criteria allow disposal of volume-contaminated animal tissue containing less than 1.85 kBq/g of ^{3}H or ^{14}C as if it were not radioactive.

[4]The committee was not able to uncover substantial evidence that this early work was based on an assessment of risk.

trol of Radioactively Contaminated Material, provides guidance on radiation control programs, including material clearance protocols (USNRC, 1981). It contains guidance for implementing the surface contamination standards in Table I, such as data on radiation detection instrumentation, as well as radiation control systems required generally of licensees. Like Regulatory Guide 1.86, this guidance is not specific to volume-contaminated materials.

Authorized Releases of Radioactive Materials from Regulatory Control—Existing and Proposed Standards

The USNRC's general radiation protection regulations in 10 CFR Part 20 prescribe acceptable radiation exposures for workers and the public, as well as permissible levels of radioactivity in gaseous or liquid emissions from licensed facilities. Section 2002 of Part 20 provides for a case-by-case review to obtain approval to dispose of radioactive materials in unlicensed facilities when procedures are not specifically prescribed by existing regulations. (The USNRC received approximately 15 such requests over the past 5 years [USNRC, 2001b]. As the committee understands it, these requests cover only proposed disposals that are different from standard practices.)

In addition to the requirements specified in Part 20, the USNRC frequently incorporates directly into a facility's license specific requirements for release of certain radioactively contaminated solid materials. Except for the exemption tables in 10 CFR Part 30, general standards for the unrestricted release of volume-contaminated solid materials have not been promulgated.

First in 1986 and again in 1990, the USNRC proposed to formalize and update the existing guidance and other regulations by establishing policies on radiation levels that would be considered "below regulatory concern" (BRC). These proposals were meant to establish a threshold for residual levels of radioactivity, below which the solid material could be cleared from further regulatory control. Section 10 of the Low-Level Radioactive Waste Policy Amendments Act of 1985 (42 U.S.C. § 2021j) specifically addresses low-level waste. Consistent with this statute, the proposed BRC policy attempted to set general criteria for allowable individual dose and collective dose[5] resulting from authorized releases of radioactively contaminated materials from licensed activities.

The BRC proposal was intended to be an overarching approach that would establish specific quantitative standards for site releases at license termination, unrestricted release of waste materials, and consumer or industrial product uses of radioactive materials, as well as other standards. In some quarters, however, this proposal was perceived as a subterfuge to reclassify a large part of the low-level waste from commercial reactors as nonradioactive waste, thereby allowing

---

[5]Collective dose is the sum of the individual doses received, in a given period of time by a specified population, from exposure to a specified source of radiation.

licensees to avoid the costs of disposal at a licensed LLRW facility (USNRC, 1991a). Many comments from the general public, the states, and Congress rejected the BRC approach for releasing radioactively contaminated materials for unrestricted reuse or disposal. In response to these criticisms, the USNRC placed a moratorium on the proposed BRC policy while it attempted to build public consensus for it. That effort failed, and Congress formally revoked the BRC policy in the Energy Policy Act of 1992. The USNRC rescinded its proposed BRC policy statement soon afterward.

In response to the USNRC's deregulation efforts, at least 16 states subsequently passed regulations or laws that were stricter than the federally proposed allowable releases. The intent evident in most of these new restrictions was to continue regulatory control if the federal government allowed deregulation. Major concerns voiced by the public included the uncertain risks, a lack of confidence in the USNRC and the Department of Energy (DOE), and general concerns about the release of radioactive materials into consumer products (USNRC, 1991a, 1991b). Chapter 8 provides further details on public reactions to the BRC proposal.

The committee has been asked to address questions related to a proposal that may be considered another attempt by the USNRC to establish uniform standards for the unrestricted release of SRSM. In 1998 the Commission directed the USNRC staff to consider a rulemaking for establishing a dose-based standard for release of SRSM (USNRC, 1998a), and in January 1999 the USNRC initiated an enhanced participatory rulemaking directed at establishing a clearance standard (USNRC, 1999c). At the same time, the USNRC sponsored a draft technical report on the topic, NUREG-1640, *Radiological Assessments for Clearance of Equipment and Materials from Nuclear Facilities* (USNRC, 1998b). This draft report was criticized severely when concerned parties learned that the contractor developing the draft, Science Applications International Corporation (SAIC), was concurrently also doing work for a company that stood to gain financially from the promulgation of a clearance standard.

The USNRC published an issues paper in the *Federal Register* (64 Federal Register 35090-35100; June 30, 1999) and held a series of public meetings from September through December 1999. Its proposal for rulemaking on release criteria aroused the same skepticism that had greeted its earlier BRC policy. Consumer and environmental groups were particularly incensed that in their view, the USNRC had predetermined the outcome before it started. These concerns led to a broad-based boycott of the first two 1999 public meetings. At the same time, the USNRC, through its contract with SAIC, was conducting a detailed technical analysis that would become NUREG-1640 to assess the risks associated with establishing a clearance standard. As discussed in the section "Stakeholder Involvement" below, significant concerns about public health and safety issues and negative economic impacts on certain industries were raised. A USNRC paper summarizing the public meetings, technical bases, and alternatives was issued on

March 23, 2000 (USNRC, 2000a). A final stakeholder briefing occurred on May 9, 2000. As part of its response to the concerns expressed at these meetings, the Commission requested that a study be undertaken by the National Academy of Sciences.

COMPARATIVE ASSESSMENT OF EXISTING REGULATIONS IN THE UNITED STATES

There are numerous regulations in the United States governing releases of radioactively contaminated materials and facilities. Three agencies—the USNRC, DOE, and EPA—have promulgated regulations and/or guidance according to their respective statutory authorities. The standards range from about 1 mrem/yr (USNRC's Regulatory Guide 1.86, as estimated in USNRC, 1998b), to 100 mrem/yr (10 CFR Part 20.1301, which limits the annual dose received by members of the public from a licensee), to 500 mrem (10 CFR Part 35.75, which allows a licensee to release a person who has received radiopharmaceuticals provided doses to *other* persons will not exceed 500 mrem). In radiation control the USNRC generally applies the standards as limits supplemented by explicit steps to maintain the exposures at levels that are as low as reasonably achievable (ALARA). The EPA generally applies specific limits to specific applications. While there is general agreement among the three agencies, differences persist with regard to standards for protection of groundwater and for an all-pathways dose. Even within one agency's regulations, there are apparent discrepancies. For instance, the cancer risks associated with EPA standards for water, air, and Superfund cleanup range over more than two orders of magnitude (NRC, 1999). In summary, the levels of protection afforded by federal regulation of radioactive materials vary widely.

USNRC Regulations

There are two sets of USNRC regulations for unrestricted release. One set pertains to the release of facilities from regulatory control; the other pertains to materials to be released on an unrestricted basis from regulated facilities. Each set of regulations provides for significant regulatory flexibility depending on the circumstances.

The USNRC's License Termination Rule: Release of Facilities

The USNRC's License Termination Rule, 10 CFR Part 20, Subpart E, governs unrestricted and restricted release of USNRC-licensed facilities from regulatory control. This rule establishes procedures and specific standards that must be met before regulatory oversight of a facility can be terminated. The rule's key requirements are as follows:

- Unrestricted release of a USNRC-licensed facility is permitted if the all-pathways dose, including groundwater, does not exceed 25 mrem/yr and radioactive residues have been reduced to levels that are ALARA.
- Restricted release of a USNRC-licensed facility is permitted if (1) the net public and environmental harm is comparable to compliance with the 25 mrem/yr limit for unrestricted release and the residue levels are ALARA; (2) institutional controls are adequately funded and legally enforceable; (3) requirements for restricted release have the advice of a broad cross section of the community interests; and (4) in the event that institutional controls fail, the maximum dose is ALARA and does not exceed 100 mrem/yr (or 500 mrem/yr under exceptional circumstances substantiated by detailed information).

Alternative criteria may be submitted by a licensee for review if they are supported by adequate plans and analyses prepared with community advice. The dose limits apply to the total effective dose equivalent for the average member of the critical group, calculated over the first 1,000 years after decommissioning.

The USNRC's Case-by-Case Approach: Release of Materials

Overview. As noted in Chapter 1, the USNRC's regulations under 10 CFR Part 20 limit the radiation dose that an individual can receive from the operation or decommissioning of a USNRC-licensed facility and also require that doses received are ALARA. Although Part 20 sets standards for releases of effluents (liquids or gases), it sets no specific standard for release of solid materials with surface or volume contamination.[6] The USNRC generally evaluates releases on a case-by-case basis using license conditions and existing regulatory guidance.

---

[6] According to the USNRC (1999a):

> For most NRC licensees, solid materials have no contamination because these licensees use sealed sources in which the radioactive material is encapsulated. These include small research and development facilities and industrial use of various devices including gauges, measuring devices, and radiography.
>
> For other licensees (including nuclear reactors, manufacturing facilities, larger educational or health care facilities, including laboratories) materials generally fall into one of three groups based on its location or use in the facility:
> - Clean or unaffected areas of a facility, from which areas the solid materials would likely have no radioactive contamination;
> - Areas where licensed radioactive material is used or stored, from which areas materials can become contaminated although the levels would likely be low to none; and
> - Material used for radioactive service in the facility or located in contaminated areas or areas where contamination can occur, from which materials generally have levels of contamination that would not allow them to be candidates for release unless they are decontaminated.

In Section 2002 of Part 20, "Method for obtaining approval of proposed disposal procedures," the basis for the case-by-case review is virtually the same as that in the old Section 302 of Part 20. As noted above, neither version provides specific standards for exemption.[7] The pertinent portion of Part 20.2002 reads as follows:

> A Licensee or applicant for a license may apply to the Commission for approval of proposed procedures, not otherwise authorized in the regulations in this chapter, to dispose of licensed material generated in the licensee's activities. Each application shall include:
> a. A description of the waste containing licensed material to be disposed of, including the physical and chemical properties important to risk evaluation, and the proposed manner and conditions of waste disposal; and
> b. An analysis and evaluation of pertinent information on the nature of the environment; and
> c. The nature and location of other potentially affected licensed and unlicensed facilities; and
> d. Analyses and procedures to ensure that doses are maintained ALARA and within dose limits in this part.

Under the case-by-case approach, the USNRC does not consider most releases of solid materials to be "disposals" authorized under Part 20 or Part 61. Instead, these releases are frequently authorized by specific license conditions, that is, a specific provision contained in the facility's license.[8]

Categories of Release. USNRC guidance on release of SRSM falls into three categories: (1) release of solid materials with surface residual radioactivity at reactors, (2) release of surface-contaminated solid materials possessed by a materials licensee (i.e. nonreactor licensee), and (3) release of volume-contaminated solid materials possessed by reactor and materials licensees (USNRC, 2001b). The guidance for each category is summarized next.

---

[7]The 1957 issue of Part 20 had a short section on waste disposal that included Part 20.302, "Method for obtaining approval of proposed disposal procedures," the basis for case-by-case review of disposal procedures not authorized by the two succeeding sections on disposal in sewerage systems or in soil. The original Part 20 gave general requirements for waste disposal of byproduct material. The 1957 standard did not include any criteria for a floor to the amount or concentration of controlled radionuclides, which criteria might be used as the basis for exemption of waste from regulatory control.

[8]It is not appropriate to apply the ALARA principle at or below the dose limits that are typically proposed for clearance calculations. These are not dose safety limits in the ordinary sense of the word, but are levels at which SRSM may be released from regulatory control. The dose limits of 0.1 to 10 mrem/yr are already orders of magnitude below natural background levels. Additionally, the variation in natural background dose is larger than the level of the selected dose limit. Since the proposed dose limits are already well below most established safety limits, it is not appropriate to apply the ALARA principle to the clearance dose limits as calculated in NUREG-1640.

Release of solid materials with surface residual radioactivity at reactors. Reactor licensees typically follow a policy established by IE Circular 81-07, *Control of Radioactively Contaminated Material*, and Information Notice 85-92, *Surveys of Wastes Before Disposal from Nuclear Reactor Facilities* (USNRC, 1981, 1985). Under this policy, reactor licensees must survey equipment and material before its release. If the survey indicates the presence of licensed AEA material above natural background levels, the equipment or material cannot be released (USNRC, 2001b). The IE Circular 81-07 and related guidance basically set the sensitivity required of survey instruments, a sensitivity similar to that used in applying Regulatory Guide 1.86.

Release of surface-contaminated solid materials possessed by a materials licensee (i.e., nonreactor licensee). For materials licensees, the USNRC usually authorizes the release of solid material through specific license conditions. Table I of Regulatory Guide 1.86 is used to evaluate surface contamination on solid materials before they are released (AEC, 1974). Similar guidance is found in Fuel Cycle Policy and Guidance Directive FC 83-23, *Guidelines for Decontamination of Facilities and Equipment Prior to Release for Unrestricted Use or Termination of Byproduct, Source or Special Nuclear Materials Licenses* (USNRC, 1983). Both documents contain a table of surface contamination criteria, which may be used by licensees as the basis for demonstrating that solid material with surface contamination can be released safely with no further regulatory control.

Release of volume-contaminated solid materials possessed by reactor and materials licensees. The USNRC has not provided guidance for volume-contaminated materials analogous to the guidance in Regulatory Guide 1.86 for surface contamination. Instead, the USNRC has decided these situations on a case-by-case basis by evaluating the doses associated with the proposed release of the material. Typically, the evaluation and decision is made in such a way as to ensure that the maximum doses are a small percentage of the Part 20 dose limit for members of the public of 100 mrem/yr.

The Role of States. Under the AEA, the USNRC has preemptive authority to license and regulate the ownership, possession, use, and transfer of AEA materials—source, byproduct, and special nuclear materials—and to set standards, as are necessary to protect public health, for the ownership, possession, use, and transfer of AEA materials. However, Section 274 of the AEA specifically authorizes the Commission to enter into agreements with states to transfer limited elements of that authority. These agreements constitute a discontinuance of USNRC's authority, not a delegation; a state assumes the USNRC's authority over selected radioactive materials (specifically, byproduct materials, source materials, or special nuclear materials in quantities not sufficient to form a critical mass). Once an agreement is signed, the USNRC continues to have an oversight responsibility to ensure that the state, called an "agreement state," has a program for the regulation of AEA material that is adequate to protect public health and safety and is compatible with USNRC regulations (USNRC, 1999b).

As of December 2001, 32 states had entered into agreements with the Commission, and four more states had applied for agreement state status. The USNRC has extensive arrangements and procedures for communicating and interacting with the agreement states, especially to ensure that agreement state regulations are compatible with USNRC regulations.

For some USNRC requirements, such as basic radiation protection standards or those that have significant implications for interstate commerce or related activity (sometimes referred to as "transboundary implications"), the agreement state must adopt essentially identical requirements, in order to be compatible with the USNRC. For other USNRC requirements, such as most licensing requirements, the agreement state has some flexibility to adopt its own requirement if the state's requirements meet the essential objective of the USNRC. States may also establish more restrictive requirements provided that they have an adequate supporting health and safety basis and the requirements do not preclude a practice that is in the national interest (USNRC, 1999b). Criteria that have been applied by states on a case-by-case basis include the use of radiation levels that are indistinguishable from background, the use of guidelines similar or equivalent to Regulatory Guide 1.86, and the use of dose-based analyses (USNRC, 1999b).

Cited Advantages and Disadvantages of the Case-by-Case Approach. The USNRC document *Control of Solid Materials: Results of Public Meetings, Status of Technical Analyses, and Recommendations for Proceeding* (USNRC, 2000a) discusses issues and concerns related to a set of alternatives for establishing control of solid materials. In particular, it summarizes the following broad advantages and disadvantages of the current case-by-case approach, an appraisal with which the committee generally agrees:

Advantages. The advantages of the case-by-case approach are the following (adapted from USNRC, 2000a):

- *It is a flexible tool that is currently in use and well understood.* The USNRC staff and licensees have developed a common understanding of the criteria involved.
- *It is protective of public health and safety.* The potential exposures received are a fraction of public health guidelines.
- *Leaving it in place would not involve additional rulemaking resources.* USNRC resources would be devoted to specific requests from licensees, which would bear the cost. The USNRC would not have to expend its resources on a rulemaking.

Disadvantages. The disadvantages of the case-by-case approach are the following (adapted from USNRC, 2000a):

- *The criteria are inconsistent and incomplete.* The absence of uniform criteria for controlling solid materials results in inconsistent release levels. Licensees also can have difficulty determining what information to provide for USNRC approval because the existing guidance and criteria may not be clear.
- *The criteria are not risk informed.* The current detection-based approach does not relate regulatory requirements to the potential risk that might be associated with the regulated activity.
- *Expenditures of time and resources are required to resolve specific cases.* Each review involves establishing and justifying criteria for that case.

DOE Standards on Clearance of Solid Materials

DOE's standards for surface contamination are set forth in Order DOE 5400.5,[9] which incorporates Table I, the surface-activity standards, from USNRC's Regulatory Guide 1.86. At about the same time as the issuance of Regulatory Guide 1.86, the regulatory staff at the AEC were asked to develop solid release standards for volume-contaminated materials from modification of the uranium enrichment plants (see Chapter 5 for a discussion of NUREG-0518). The development of that standard was set aside after publication of NUREG-0518 (USNRC, 1980). Since then, DOE has maintained a policy that generally precluded the release of radioactively contaminated materials for unrestricted use or disposal. Not until Assistant Secretary of Environmental Management Alvin Alm issued a policy statement in September 1996 promoting, on a provisional basis, the recycling of radioactively contaminated scrap steel did DOE formally alter its long-standing policy against unrestricted release of contaminated materials. DOE's release policy had initially focused narrowly on restricted end uses of recycled steel at DOE facilities. It was subsequently broadened, at least unofficially, to include recycling into industrial and consumer products generally.

The 1996 policy change was implemented on a conditional basis while DOE evaluated the safety and economics of recycling these materials. The first large-scale project involving the recycling of radioactively contaminated materials was initiated at the Oak Ridge Reservation's gaseous diffusion plants, which contain

---

[9]Order DOE 5400.5, *Radiation Protection of the Public and Environment*, Department of Energy, February 8, 1990, revised January 7, 1992. A DOE memorandum dated November 17, 1995, from R.F. Pelletier, provided field and program offices with additional guidance regarding control of residual radioactive material, including the relationship of DOE standards to similar standards set by the USNRC and the states.

more than 100,000 tons of contaminated metals (EPA, 1997a; NRC, 1996). The Oak Ridge project was intended to establish a precedent for a much broader reliance on reuse of radioactively contaminated materials throughout the nuclear weapons complex (NRC, 1996). Under current estimates, DOE facilities contain about 1 million tons of contaminated metals that could be recycled (EPA, 1997a).

Contrary to the recommendations of a prior National Research Council report (NRC, 1996), the Oak Ridge project proceeded with little public outreach, and it ultimately provoked significant opposition from the public and the metals recycling industry. In response to this strong opposition from both the private sector and the public, Secretary of Energy Bill Richardson halted further releases of volume-contaminated metals—but not surface-contaminated metals—from DOE facilities in January 2000. The moratorium was limited to volume-contaminated metals because no generally accepted regulatory standard or guidance existed. In July 2000, Secretary Richardson reaffirmed this moratorium on volume-contaminated materials and added a temporary suspension on unrestricted recycling of all scrap metal originating from within radiologically controlled areas. He proposed continuing the moratorium and suspension until the USNRC resolved whether to proceed with promulgating a standard governing the clearance of radioactively contaminated solid materials. DOE, however, recently initiated the process for drafting a programmatic environmental impact statement on alternatives for recycling surface-contaminated metals (DOE, 2001).

The EPA Role

Under the AEA and Reorganization Plan No. 3 of 1970, the EPA has responsibility for establishing radiation standards, with which USNRC's and DOE's standards must conform. EPA has used its AEA authority to promulgate standards such as 40 CFR Part 190, which sets limits on doses received by members of the public from nuclear power operations. Pursuant to other statutes, EPA has promulgated radiation standards for air emissions and safe drinking water levels. Under the Safe Drinking Water Act, the EPA established a 4 mrem/yr standard for the dose that an individual is permitted to receive from drinking water (40 CFR Parts 141-142). This standard is based on a single pathway of exposure, under which an individual consumes 2 liters of water per day from a single source of drinking water. Under the Clean Air Act, the EPA promulgated the National Emission Standards for Hazardous Air Pollutants (NESHAP), which permits a 10 mrem/yr dose to the reasonably maximally exposed individual from airborne emission of radioactive materials (40 CFR Part 61). The basis for this standard includes multiple exposure pathways, including exposure from airborne plumes, inhalation, and ingestion of foods on which radioactive materials have been deposited.

In August 1997 the EPA issued guidance on residual levels of radionuclides permitted under the Comprehensive Environmental Response, Compensation, and Liability Act (CERCLA) (EPA, 1997b). The agency premised its standard on the policy that remediation goals for radionuclides should be consistent with a lifetime risk ranging from 10^{-4} to 10^{-6}. According to EPA guidance, clearance levels for CERCLA sites cannot result in a dose that exceeds 15 mrem/yr, which EPA guidance states "equates to approximately 3×10^{-4} increased lifetime risk" (EPA, 1997b).

In the context of evaluating potential clearance, or de minimis, standards, the EPA has provided technical analyses in the form of two major studies. In 1997 it completed a draft technical support document, *Evaluation of the Potential for Recycling of Scrap Metals from Nuclear Facilities* (EPA, 1997a), and a cost-benefit analysis, *Radiation Protection Standards for Scrap Metal: Preliminary Cost-Benefit Analysis* (EPA, 1997c).

The focus of EPA standard setting for unrestricted release has been on promoting consistent international import-export controls for materials containing residual radioactivity. This issue has become increasingly important with the erosion of regulatory controls at nuclear facilities in the countries of the former Soviet Union. A number of incidents have occurred in the United States and elsewhere in which radioactive materials have been discovered in scrap metal loads at steel mills and, less frequently, have contaminated the metal used to fabricate consumer products as in the Ciudad Juarez, Mexico, incident in 1983 (Lubenau, 1998).

In 1998 the EPA began to work with the International Atomic Energy Agency (IAEA) on clearance issues and import-export standards. EPA personnel initially worked on technical issues in an effort to promote agreement between the parties on appropriate methodologies for estimating exposure levels.

Control of TENORM

Naturally occurring radionuclides are found throughout the United States, primarily in the form of elements such as uranium, thorium, radium, potassium, and radon gas (NRC, 1999). Industrial activities such as oil and gas extraction, water treatment, mining, fossil fuel processing, and aluminum production generate tens of billions of metric tons of TENORM, some of which contain high levels of radioactivity (NRC, 1999). However, TENORM is not subject to the AEA because it cannot be classified as a source material, special nuclear material, or byproduct material.

Federal regulation of TENORM has been largely absent. In 1986 the Radon Gas and Indoor Air Quality Act directed the EPA to study the dangers of TENORM, particularly radon gas. After completing this study, the EPA drafted proposed rules to regulate TENORM under the Toxic Substances Control Act, which gives EPA the authority to regulate chemical substances, including those that are naturally occurring, that may present an "unreasonable risk of injury to health or the environment" (EPA, 1989). The EPA's draft proposed rules were stayed indefinitely. An exception to this void in regulating TENORM is Order DOE 5400.5, which DOE issued under its general responsibility to protect health and safety in conducting activities authorized under the AEA.

This regulatory gap persists despite the fact that many forms of TENORM can be substantially more radioactive than LLRW subject to regulation under the AEA (NRC, 1999). The existing state regulations that apply to TENORM have largely been limited to disposal and handling requirements enacted under the state's general radiation protection laws or under other authority, such as the Resource Conservation and Recovery Act. The Conference of Radiation Control Program Directors (CRCPD) has drafted model state regulations for TENORM, but these have been neither finalized nor adopted by any states. State regulations remain limited and vary greatly from state to state (CRCPD, 1997).

STAKEHOLDER INVOLVEMENT

As noted earlier, the current evaluation of clearance of solid materials by the USNRC is not the first time it has attempted to update and formalize guidance for unrestricted releases of SRSM. The most notable prior attempts were those in 1986 and 1990 (discussed above) to establish policy and guidance for solid materials whose residual radioactivity would be "below regulatory concern." These attempts and the subsequent stakeholder reactions provide invaluable insight into the current USNRC effort to establish uniform standards for release of SRSM.

After the 1990 BRC policy statement was published in the *Federal Register* (55 Federal Register 27522; July 3, 1990), the USNRC held public meetings in five cities (USNRC, 1991a). These meetings were contentious and well attended by representatives of a large number of stakeholder groups. The USNRC estimated that more than 900 people attended, and oral statements were taken from 215 people. The oral statements were supplemented by numerous written questions and comments. "The prevailing sentiment expressed at each of the meetings was one of opposition to the BRC policy and to its implementation" (USNRC, 1991a).

In 1991 the USNRC staff reported that three themes were common to the five public meetings. First, "extreme concern was expressed concerning the possibility of deregulation of nuclear power waste." Second, many attendees from the public (including a large number of environmental groups) stated their strong opposition to recycling of materials that could be used in unlabeled consumer products. Third, many attendees perceived that the policy would "permit a large number of deaths per year per practice despite the presence of collective dose criterion" (USNRC, 1991a). In short, stakeholders' concerns expressed at the meetings centered on whether the USNRC could adequately protect the public.

Many of these stakeholders also expressed the belief that low levels of radiation were much more harmful than the regulatory agencies had determined them to be. This expressed fear was compounded by concerns that it would not be possible to monitor solid materials adequately for radioactivity when they were being surveyed before release. Many of the stakeholders also raised two closely related issues. First, many alleged that the regulatory system failed to take into account multiple exposures. Second, general standards for release would undermine individual rights to decide the nature and magnitude of the risks to which members of the public would be exposed. These issues continue to be central to stakeholder criticisms. Most of the stakeholder concerns still revolve around safety and protection of the public.

The nuclear industry strongly supported the 1990 BRC policy, as did a few other stakeholder groups, on the grounds of economic and resource efficiency. However, the sheer number of groups opposing the policy; the intensity of their viewpoints; and their consistency in raising issues of public health, safety, and welfare doomed this policy from the outset. After the policy was announced in 1990, the USNRC hired a consultant to begin a phased consensus-seeking process. This effort collapsed shortly after it started because public interest groups refused to engage in the process (USNRC, 1991b). As noted above, Congress formally nullified the BRC policy as part of the Energy Policy Act of 1992. Even before Congress acted, the USNRC issued a moratorium on the BRC policy in July 1991 (56 Federal Register 36068-36069; July 30, 1991); after the Energy Policy Act was signed into law, the USNRC rescinded the policy in August 1993.

The BRC policy was defeated largely by the efforts of these public interest groups, which successfully used the political arena to expand the controversy over the issue and to make the issue salient to a large number of stakeholder groups and other interested parties.

The lines that were drawn in 1991 over the BRC policy do not seem to have altered appreciably. Many of the public interest groups that the USNRC concluded were indispensable to any effort to promote a consensus-seeking process are adamantly opposed to the proposed USNRC rulemaking on SRSM. The only shifts that have occurred are in the positions of officials from several states, whose representatives had opposed the BRC policy solely because of concerns that it would abrogate the states' enforcement authority. However, even among the agreement states from which the committee has heard, there is no consensus on the proposed rule. Many of those who addressed the committee questioned whether such a rule is necessary at all. What lessons, if any, the USNRC has learned from the BRC controversy is a question that the committee addresses in Chapter 9.

FINDINGS

Finding 2.1. The USNRC does not have a clear, overarching policy statement for management and disposition of SRSM. However, SRSM has been released from licensed facilities into general commerce or landfill disposal for many years pursuant to existing guidelines (e.g., Regulatory Guide 1.86) and/or following case-by-case reviews. The USNRC advised the committee of no database for these releases.

Finding 2.2. A dose-based clearance standard can be linked to the estimated risk to an individual in a critical group from the release of SRSM. The general regulatory trend is toward standards that are explicitly grounded in estimating risks.

Finding 2.3. For clearance of surface-contaminated solid materials, the clearance practices regulated by the USNRC and agreement states are based on the guidance document Regulatory Guide 1.86, which is technology based and has been used satisfactorily in the absence of a complete standard since 1974.

Finding 2.4. For clearance of volume-contaminated solid materials, the USNRC has no specific standards in guidance or regulations. Volume-contaminated SRSM is evaluated for clearance on a case-by-case basis. This case-by-case approach is flexible, but it is limited by outdated, incomplete guidance, which may lead to determinations that are inconsistent.

Finding 2.5. Industrial activities are generating very large quantities of technologically enhanced naturally occurring materials (TENORM). Federal regulation of TENORM has been largely absent. State regulations vary in breadth and depth.

3

Anticipated Inventories of Radioactive or Radioactively Contaminated Materials

This chapter summarizes current estimates of the quantities of slightly radioactive solid material (SRSM) expected to arise over the next 25 years from cleanup and decommissioning of licensed nuclear facilities and from other facilities that may contain SRSM. These estimated inventories include materials from U.S. Nuclear Regulatory Commission (USNRC)-licensed facilities, from facilities licensed by agreement states, and from U.S. Department of Energy (DOE) and Department of Defense (DoD) facilities that do not require a USNRC license. Radioactively contaminated materials known as naturally occurring radioactive material (NORM), naturally occurring and accelerator-produced radioactive material (NARM), or technologically enhanced NORM (TENORM) also arise from a variety of activities that are not subject to the Atomic Energy Act (AEA) and thus are not regulated by the USNRC. The latter materials are not federally regulated but are regulated by state agencies in some states or not regulated at all in other states. Thus, the USNRC needs to be aware that any new regulations regarding clearance of SRSM could also have impacts on the management of contaminated materials that are currently unregulated at the federal level. Some perspective is also provided in this chapter on the relative fraction of the annual amount of recycled commercial steel scrap that cleared SRSM could comprise if clearance for unrestricted recycle were to be approved.

The committee did not find readily available information on inventory and anticipated dates for disposition of radioactive materials. The information currently available covers some industries but not others. In some cases, inventories of radioactive materials have been developed based on what is currently being generated from active licensed operations. Other inventories have been developed based on projections of future decommissionings.

Inventories for materials that fall outside the legal requirements for radioactive waste management are not as carefully developed. The unlicensed industry segments, such as many that produce NORM or TENORM, deal with radioactive material as an unwanted byproduct associated with industrial processes. Inventory information about NORM and TENORM tends to focus on the concentrations of radium, uranium, or thorium and daughter radionuclides that they contain, rather than on total inventories.

Therefore, one must often infer or estimate the amount of materials that may satisfy particular clearance criteria based on information created for a different purpose. This chapter relies heavily on a recent report *Inventory of Materials with Very Low Levels of Radioactivity Potentially Clearable from Various Types of Facilities*, which was prepared for the USNRC by Sanford Cohen & Associates, Inc. (SCA, 2001). Information from this source has been supplemented with information from various published and Internet sources and from materials presented to the study committee.

The characteristics and quantities of radioactive materials used or possessed by USNRC licensees are discussed in the following section. To provide the bases for the cost analysis given in Chapter 4, the emphasis in that section is on radioactive material streams arising from the decommissioning of licensed power reactors. To complete the picture of radioactive materials in the United States, summary information on the other licensed and unlicensed radioactive material streams is presented in the second section.

INVENTORIES OF CONTAMINATED MATERIALS ARISING FROM DECOMMISSIONING OF USNRC-LICENSED FACILITIES

The majority of USNRC-licensed facilities can be divided into four types, each of which produces a characteristic body of radioactive materials during operations and decommissioning: (1) nuclear reactors (electric power, materials testing, and research reactors); (2) fuel cycle facilities (uranium milling, UF_6 [uranium hexafluoride] conversion plants, and uranium fuel fabrication); (3) non-fuel-cycle facilities (radioactive material processing, research laboratories, medical treatment, radiography, etc.); and (4) independent spent fuel storage installations (ISFSIs), which store spent fuel from power reactor operations.

Because of the substantial number (more than 100) and large size of electric power reactors, they are the source of about 75 percent of the radioactive materials in the United States that require disposal in licensed low-level radioactive waste (LLRW) disposal sites. Power reactors also provide SRSM that is cleared from regulatory control. SRSM arising from the latter three types of facilities is examined in less detail in this report because the quantities of radioactive materials arising during operation or during decommissioning are small compared to the quantities arising from power reactor decommissioning.

Power Reactors

Some data are available for estimating the types and annual quantities of radioactive materials arising from the operation of power reactor facilities that currently dispose of their LLRW at licensed LLRW disposal facilities. Additional data and various estimates are available to define the types and total quantities of radioactive materials resulting from decommissioning power reactor facilities. The decommissioning data and estimates presented in Table 3-1 are derived from two USNRC reports: NUREG/CR-5884 (Konzek et al., 1995) for a reference pressurized water reactor (PWR) and NUREG/CR-6174 (Smith et al., 1996) for a boiling water reactor (BWR). Also presented in the table are estimates of the sums of the quantities of these materials expected to arise from the total U.S. population of power reactors. These population estimates were scaled from the reference reactor quantities using multiplication factors derived from the SCA (2001) report on inventory using the following equations:

$$M_{Pop.P} = M_{Ref.P}\, \Sigma i\, (P_{Pi}/P_{Ref.P})^{2/3}$$

and

$$M_{Pop.B} = M_{Ref.B}\, \Sigma i\, (P_{Bi}/P_{Ref.B})^{2/3}$$

where $M_{Pop.P}$ and $M_{Pop.B}$ are the PWR and BWR population multipliers, respectively, $M_{Ref.P}$ and $M_{Ref.B}$ are the weights of radioactive materials postulated to arise from decommissioning the reference PWR and BWR, respectively; $P_{Ref.P}$ and $P_{Ref.B}$ are the rated power levels of the reference PWR and BWR, respectively, and P_{Pi} and P_{Bi} are the rated power levels of the individual PWRs and BWRs that make up the U.S. population of power reactors. In essence, the population multiplier for a PWR or BWR represents the number of reference PWRs or BWRs that would contain the same total amount of structural material as is contained within the total populations of PWRs and BWRs that exist currently in the United States. Because many of the reactors are smaller than the reference reactors, the population multipliers are smaller than the actual number of each type of reactor in the total population.

For this analysis, the total volume of potential LLRW estimated to arise from decommissioning a power reactor is divided into three categories: (1) activated materials,[1] including the reactor pressure vessel and internals and the activated portions of the biological shield; (2) nonreusable contaminated materials such as

---

[1] Materials made radioactive through irradiation of stable nuclides by neutrons, protons, electrons, or other particles or radiation.

TABLE 3-1 Volume of Materials Arising from Power Reactor Decommissioning (cubic meters)

Material Type	PWR Volumes[a]	BWR Volume[b]	Population Totals
Activated (LLRW)	547	889	60,900
Nonclearable (LLRW)	1,800	1,520	159,000
Metallic SRSM	5,830	12,700	743,000
Excluded (30%) as LLRW	1,750	3,820	233,000
Net SRSM	4,080	8,900	521,000
Concrete SRSM	69,500	99,700	7,360,000
Total volumes SRSM	73,600	109,000	7,880,000
Population multipliers[c]	63.76	29.23	

NOTE: All values are rounded to three significant figures.

[a]Konzek et al. (1995).
[b]Smith et al. (1996).
[c]Data derived from SCA (2001). Each multiplier represents the number of reference reactors of that type that would contain the same total amount of structural material as is contained within the total population of each reactor type.

ion-exchange resins, filters, plastics, contaminated equipment insulation, and removed contaminated concrete surfaces; and (3) metallic SRSM that might be uncontaminated but is from a radioactive work area or that might be only slightly contaminated. The metallic SRSM includes pool liners, piping, tanks, valves, pumps, heat exchangers, and similar items. Because of the complexity of their inner and outer surfaces, it is difficult to demonstrate that some of these items (such as heat exchangers, pumps, and valves) have been decontaminated sufficiently to permit release under a clearance standard. An examination of the tables of system components presented in Konzek et al. (1995) shows that roughly 30 percent of the volume of the metallic SRSM in those tables would probably be excluded on the basis of structural complexity. For this analysis, that 30 percent fraction has been excluded from the volume of SRSM and equipment when calculating the volumes in Table 3-1. The same fraction was assumed to be applicable to the metallic SRSM arising from decommissioning a BWR.

The structural concrete rubble arising from demolition of decontaminated facility structures (clearable concrete) represents the largest single component of the decommissioning wastes. The volumes presented in the table are, for the purposes of analysis, based on the assumption that after contaminated surfaces and activated concrete have been removed, the remaining concrete structures are essentially uncontaminated and may be suitable for clearance or conditional clearance (e.g., for reuse in highway construction or other uses, or for disposal in municipal waste Resource Conservation and Recovery Act [RCRA] Subtitle D landfills). The volumes of concrete SRSM rubble are larger than the combined

TABLE 3-2 Weights of Slightly Radioactive Solid Material from Power Reactors (metric tons)

Material Type	PWR Weights	BWR Weights	Population Totals
Metallic SRSM	7,860	18,700	1,050,000
Excluded as LLRW (30%)	2,360	5,610	315,000
Net metallic SRSM	5,500	13,100	735,000
SRSM concrete[a]	83,600	120,000	8,850,000
Total weight SRSM	89,100	133,000	9,590,000

NOTE: Values are rounded to three significant figures and were derived from Konzek et al. (1995) and Smith et al. (1996).

[a]From Table 3-1, by assuming that the density of concrete rubble is 1.2 metric tons per cubic meter.

volumes of all of the other SRSM by at least a factor of 10. Although it is assumed that beyond the surface, the remainder of the concrete is uncontaminated, determining what to do with the concrete is complicated by several factors. It can be difficult, in practice, to determine the quantities and levels of radionuclide contamination that have penetrated into the concrete. There are also sampling and analysis costs associated with demonstrating that material is clean, as discussed in Chapter 6 in Measurement Cost. Public perception and regulatory factors can affect choices a licensee makes on disposition of such material, such as whether concrete is left as on-site fill after the license of a site is terminated. The committee was informed that these difficulties with on-site disposal have been encountered with at least one decommissioning of a reactor site, Maine-Yankee.

Table 3-2 presents the weights of SRSM and clearable concrete estimated from the reference PWR and BWR. Population totals assume that the same population-scaling factors applied to material volumes in Table 3-1 also apply to material weights.

The time distribution of these decommissioning wastes is a significant consideration. The quantities of material arising from decommissioning nuclear power reactors will be distributed over an extended period because of the varying dates at which their licenses are scheduled to expire (SCA, 2001, Tables 2-26, 2-27). Figure 3-1 illustrates this time distribution for the weight of metallic and concrete SRSM, given the shutdown dates stated in SCA (2001). If licenses are extended for an additional 20 years, which seems probable for most facilities, the large quantities of material shown in the figure would be generated up to 20 years later, with little material resulting from decommissioning until after 2030.

With or without license extensions, the weights of decommissioning material requiring disposition (about 8 percent metals and 92 percent concrete) range from about 100,000 to more than 1 million metric tons per year during a 25-year

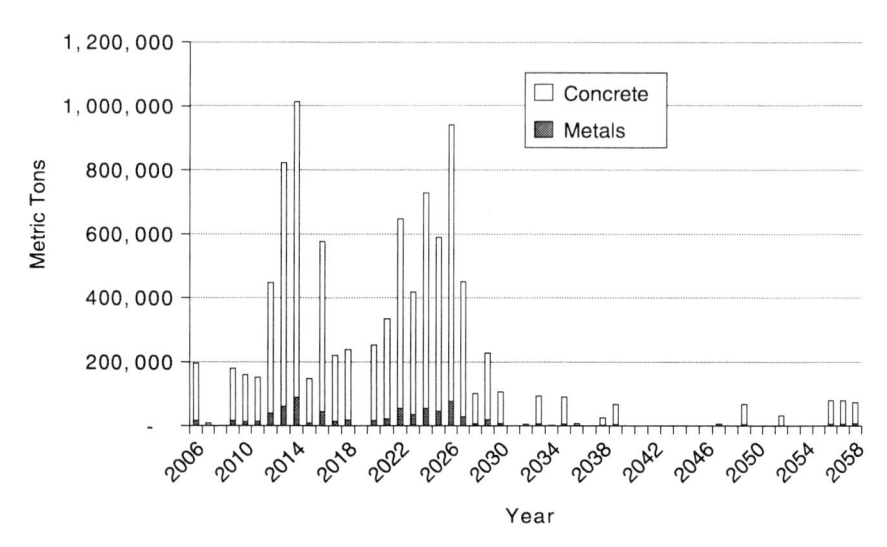

FIGURE 3-1 Time distribution for generation of slightly radioactive solid material from U.S. power reactor decommissionings. SOURCE: Adapted from SCA (2001).

period. The average is around 360,000 metric tons per year, or the equivalent of decommissioning four or five power reactor units per year. If most of the currently operating reactors do receive 20-year license extensions and if the reactors already in safe storage are decommissioned as assumed in SCA (2001), most of the weights shown in Figure 3-1 between 2006 and 2030 would move roughly 20 years into the future, to 2026 to 2050. Relatively small quantities of SRSM from power reactor decommissioning would be generated during the next three decades.

It is instructive to compare the amount of ferrous metals arising from decommissioning activities at commercial power reactors with the total amount of ferrous metal scrap currently being recycled commercially. The committee heard from a representative of a major scrap broker-processor[2] that the average amount of obsolete scrap recycled into commercial steelmaking in the years 1997-1999 was about 42 million metric tons per year. During the same period, U.S. production was about 98 million metric tons per year. The amount of nonactivated, steel SRSM arising from decommissioning the population of U.S. power reactors, as shown in Table 3-2, ranges from 0.74 million to 1.05 million metric tons (depending on the amount excluded as LLRW). Based on the distribution of current license expiration dates for U.S. power reactors over a 25-year period, the average amount of steel SRSM would be between 30,000 and 42,000 metric tons per

---

[2]Presentation to the committee by Ray Turner, David J. Joseph Company, June 13, 2001, Washington, D.C.

year. If the larger quantity (42,000 metric tons per year) was recycled, the potentially radioactive scrap would constitute only about 0.1 percent of the total steel scrap recycled each year. This small amount of metallic SRSM indicates that the effect on the available scrap metal resources is negligible if the metal is not recycled.

Nonpower Reactors

There are 46 USNRC-licensed research reactors in the United States, of which 36 are still operational (SCA, 2001, Table 2-79). Konzek et al. (1995) developed a decommissioning materials inventory for a reference research reactor that is presented again in SCA (2001). Also given in SCA (2001) are decommissioning data from four retired research reactors. The data from these four reactors were used in a least-squares analysis to develop a scaling factor for the weight of decommissioning material as a function of the licensed power rating of each research reactor relative to the reference research reactor (SCA, 2001, p. 2-138). The resulting equation for the scaling factor is $M_i/M_R = [P_i/P_R]^{1.0813}$, where M is the weight of material and P is the power rating, for the ith reactor and the reference reactor, respectively. The R^2 value for the fit of the data to the equation was 0.97.

The power ratings for the four research reactors used in the analysis ranged from 5 W to 20 MW, and the power rating of the reference research reactor was just 1.1 MW. Because a certain amount of facility structure is needed almost regardless of the power rating of the contained reactor, this scaling factor may underestimate the quantities of materials arising from research reactors having the much lower power ratings. Computing this factor for each of the 46 licensed research reactors and summing over that population yields the population scaling factor (65.79). Multiplying the weights of each category of materials (structural steel, concrete, system steel) from the reference research reactor by the population scaling factor yields the population weights for each material category from U.S. research reactors, as shown in Table 3-3. The weights of structural steel and concrete SRSM are assumed to all be clearable, without any exclusions for LLRW materials. The study committee also assumed that metallic SRSM from the system steel category would have the same 30 percent fraction that would have to be disposed of as LLRW as assumed in the previous section on power reactors. The inventory of steel and concrete from research reactors represents about 1.4 percent of the total weight of SRSM from the power reactors.

INVENTORIES OF RADIOACTIVE WASTE FROM OTHER LICENSED AND UNLICENSED SOURCES

Radioactive materials are generated in a number of industrial environments, where the sources range from dilute to concentrated and from small volumes to

TABLE 3-3 Decommissioning Materials Inventory from the Population of U.S. Research Reactors (metric tons)

Composite Reactor	Structural Steel	Concrete	System Steel
Activated	—	—	6.5
Nonclearable	—	11	2.0
SRSM	113	1,910	46.0
Excluded (30%)	—	—	13.8
Net SRSM	113	2,010	39.9
Population weight SRSM	7,400	125,000	2,100

NOTE: Values are rounded to three significant figures. Population scaling factor is 65.79.
SOURCE: Data derived from SCA (2001).

large volumes. The information presented here is intended to provide a broad view of the types and quantities of radioactive materials present in the United States. Some of these materials are under federal regulatory control, others are under the control of state agencies, and still others may not be under any regulatory control. The inventories include radioactive materials generated by (1) fuel cycle and (2) non-fuel cycle facilities, both categories of which are licensed, permitted, and regulated by the USNRC and agreement states; (3) facilities subject to the USNRC's Site Decommissioning Management Plan (SDMP); (4) DOE facilities; (5) DoD facilities; (6) facilities regulated by the Environmental Protection Agency (EPA Superfund sites) or state agencies; and (7) industries that produce NORM, NARM, or TENORM.

Steel and concrete SRSM arise from decommissioning activities at fuel-cycle and non-fuel-cycle facilities. The SRSM generated at these sites will include some or all of the following:

- Surface-contaminated equipment and material (i.e., concrete), and
- Materials that are not from controlled radioactive areas and may be designated as clearable, depending upon the type of facility.

In general, activated metals and concrete have been and will continue to be disposed at licensed LLRW disposal facilities. These activated materials are not considered candidates for clearance, except where the concentration of activation products is very minimal. The category of surface-contaminated equipment and material includes some materials that are unlikely to be clearable and some that might be clearable after application of an appropriate decontamination technology. The types and quantities of radioactive materials arising from decommissioning each type of facility are discussed briefly below.

USNRC-Licensed Fuel Cycle Facilities

There are basically four types of fuel cycle facilities licensed by the USNRC: uranium mills, uranium hexafluoride conversion plants, uranium oxide fuel fabrication plants, and ISFSIs.

Uranium Mills

The population of uranium mills consists of four conventional surface ore crushing and/or leaching facilities and up to seven (one is not yet operational) in situ leaching facilities. In the surface mills, the waste materials from decommissioning are generally disposed by adding them to the ore tailings piles. Little waste remains that would require disposal at an LLRW facility. The in situ leaching facilities produce some wastes for LLRW disposal, and some of their surface structures and equipment may be conditionally clearable. The contaminants present are primarily natural uranium (^{235}U and ^{238}U and their daughter products). No data are readily available on the volumes and weights of material and equipment that will arise from decommissioning in situ leaching facilities. However, because of the simplicity of these facilities, the committee expects that the quantities will be small.

Uranium Hexafluoride Conversion Plants

Decommissioning of the two existing uranium hexafluoride conversion plants is expected to be completed ultimately. One is currently operating; the other has been undergoing decommissioning for the past eight years. Although these two plants use different chemical processes, the SCA (2001) report assumes that they are sufficiently similar that a scaling factor of 2 is appropriate for calculating the size of the population waste inventory. The anticipated contaminants are primarily natural uranium (^{235}U and ^{238}U and their daughter products), with concentrations in the range of 10 to 10,000 pCi/g. Table 3-4 gives the estimated weights of radioactive materials arising from decommissioning these facilities. For the uncleared equipment, the study committee accepted the assumption made by Elder (1981) that 40 percent is LLRW and 60 percent is SRSM. For the non-LLRW concrete and structural steel (including reinforcing bar in concrete, or rebar), Elder (1981) assumed that 40 percent is SRSM and 60 percent is clearable. Because there are only two of these facilities, the quantities requiring disposition are small.

Uranium Fuel Fabrication Facilities

There are seven uranium fuel fabrication plants presently licensed in the United States. Their licenses are currently scheduled to expire 2001 to 2009. At

TABLE 3-4 Decommissioning Materials Inventory from the Population of U.S. Uranium Hexafluoride Conversion Plants (metric ton)

Materials	Structural Steel	Concrete	Equipment
LLRW	—	161	928
SRSM	616	3,250	1,390
Clearable	922	4,870	271
Total clearable	1,540	8,120	1,660

NOTE: Values are rounded to three significant figures.
SOURCE: Data are derived from SCA (2001).

least four of these plants will probably have their licenses extended, in order to serve the U.S. nuclear power industry and the nuclear navy. Thus, the material inventories arising from decommissioning the population of uranium fuel fabrication plants, shown in Table 3-5, are likely to be distributed over the next 30 years or more.

The principal contaminants are low-enriched uranium (^{235}U and ^{238}U and their daughter products). The radioactivity levels on plant equipment could range from essentially zero up to 38,000 pCi/g.

For the committee's analysis, only six of the seven plants were considered; the naval reactors fuel plant was omitted. Table 3-5 uses a committee-derived population scaling factor, developed using the formula in SCA (2001), for estimating the weights of materials in other plants from the weights in a reference fuel fabrication plant (Wilmington, North Carolina), for which data were given in SCA (2001). For equipment, the same assumptions were used that were made for

TABLE 3-5 Decommissioning Materials Inventory from the Population of U.S. Fuel Fabrication Plants (metric tons)

Materials	Structural Steel	Concrete	Equipment
LLRW	347	2,010	
SRSM	6,500	21,000	3,020
Clearable	9,750	31,500	4,400
Total clearable	16,300	52,500	7,420

NOTE: The committee used a scaling factor of 3.88 applied to the reference plant value. Values are rounded to three significant figures.
SOURCE: Reference plant data are from SCA (2001).

the uranium hexafluoride plants. Namely, of the uncleared material, 40 percent would be disposed as LLRW and 60 percent is SRSM. For concrete and structural steel (including rebar), 40 percent is assumed to be SRSM and 60 percent is assumed clearable.

Independent Spent Fuel Storage Installations

An independent spent fuel storage installation (ISFSI) is a facility in which spent nuclear fuel from a nuclear power reactor is stored, primarily fuel that is in excess of the capacity of the spent fuel pool at the reactor. There are 15 ISFSI facilities in service in the United States employing five design concepts:

1. Vertical ventilated concrete casks (four sites),
2. Horizontal storage modules (eight sites),
3. Vertical metal casks (one site),
4. Modular vault dry storage (one site), and
5. Water-filled pool (one site).

Additional facilities are planned to be constructed in the coming decade to accommodate the excess spent fuel accumulating at reactors until a federal deep geologic repository begins receiving spent fuel for disposal.

The interior surfaces of the metal storage canisters in the dry storage concepts will undoubtedly be contaminated and might actually be activated to very low activity levels. However, the quantities of SRSM are not large and would accumulate slowly. The accumulation rate will be determined by the rate at which the geologic repository receives spent fuel. Thus, the committee has concluded that these materials will not contribute significantly to the total quantity of materials entering the disposal stream during any given year.

Non-Fuel-Cycle Licensees of the USNRC or Agreement States

There were roughly 21,000 radioactive materials licensees in the United States in 2000, consisting of roughly 5,000 USNRC licensees and nearly 16,000 agreement state licensees. Of the various types of licensees in this group, those involved in research and development, medical applications, nuclear pharmaceuticals, and the manufacture of sealed sources and radio-labeled compounds generate materials potentially subject to a clearance regulation. The estimates for radioactively contaminated materials generated by these licensees were calculated by multiplying the estimated weight of SRSM in a reference facility by the number of USNRC-licensed facilities of the same type. This result was then multiplied by 4 to account for the 75 percent of radioactive materials licenses issued by agreement states (SCA, 2001).

Hospitals

SRSM in hospitals consists of floors, walls, equipment (metal), and cabinets (wood). The total U.S. inventory is approximately 436,000 metric tons, of which an estimated 8,720 to 21,800 metric tons is disposed annually. Most of these materials are clearable. However, some small percentage contains fixed ^{3}H and ^{14}C contamination that must be disposed of as biomedical LLRW.

Research and Development Laboratories

The inventory of possibly radioactive materials in the reference research and development laboratory was estimated in SCA (2001) to be about 1 metric ton of equipment and about 2.5 metric tons of concrete. Hot cells and fume hoods were not included in the estimates, since they are expected to contain too much contamination to be considered for clearance. The total U.S. inventory for research and development laboratories was estimated by SCA (2001) to be about 2,058 and 5,145 metric tons of equipment and concrete, respectively.

Manufacturers of Sealed Sources and Radio-Labeled Compounds

Manufacturers of sealed sources and radio-labeled compounds use licensed radioactive materials in hot cell laboratories. Potentially clearable materials consist of approximately 1.7 metric tons of metal, concrete, and asphalt tiles in the reference facility, or about 107 metric tons for the 63 such facilities in the United States (SCA, 2001).

Biomedical Wastes

Biomedical radioactive waste is generated under either USNRC or agreement state licenses by institutions engaging in medical, biological, or academic research and in universities and hospitals where radioactive materials are used for research, diagnosis, or treatment of disease. Biomedical use of radioactive materials typically generates small volumes of LLRW with low content of radioactivity. Although short-lived radionuclides are most often used in biomedical research, longer-lived radionuclides such as tritium and ^{14}C are also used.[3] The longer-lived wastes are disposed at licensed LLRW facilities after pretreatment to reduce waste volume, which reduces disposal costs. Much of the short-lived waste can be managed by storage for decay, with subsequent disposal according to the nonradioactive constituents of the wastes (NRC, 2001).

---

[3]Criteria in 10 CFR Part 20 allow disposal of volume-contaminated animal tissue containing less than 1.85 kBq/g of ^{3}H or ^{14}C as if it were not radioactive.

Facilities Under the Site Decommissioning Management Plan

The USNRC is regulating the decommissioning of 28 facilities under the SDMP. Radioactive residues at these facilities consist primarily of ore or slag containing elevated concentrations of natural radioactivity (i.e., uranium and thorium and their daughter products). Approximately 4,100 cubic meters (9,840 metric tons) of concrete SRSM is expected to be produced. About 84,000 cubic meters of slag from previous processes may be recovered for reprocessing or other controlled uses.

DOE Facilities

Numerous DOE facilities have moved from production to decontamination and decommissioning. Assuming that 25 percent of the steel and iron present at these facilities cannot be recycled for economic or radiological reasons, recent studies estimate that about 1 million metric tons of metallic SRSM exist in current inventory or are expected to become available by 2035 (SCA, 2001). An estimated 60 percent of these metals will come from decommissioning the gaseous diffusion plants located at Oak Ridge, Tennessee (the K-25 plant); Piketon, Ohio ("Portsmouth"); and Paducah, Kentucky. The radionuclides of concern at the gaseous diffusion plants include ^{235}U, ^{238}U, ^{239}Pu, ^{237}Np, and ^{99}Tc. Concentrations tend to be dilute, with 78 percent of the ferrous metals estimated to contain less than 4,400 Bq/kg (120 pCi/g). (The significance of these concentrations depends on the scenarios whereby the radionuclides could expose humans to a radiation dose. This issue is covered in detail in Chapter 5.)

As discussed in the section on decommissioning power reactors, the amount of steel scrap recycled into commercial steelmaking is currently about 42 million metric tons per year. The projected 1 million metric tons of steel SRSM generated from DOE decommissioning and cleanup operations are expected to become available over about a 25-year period, or an average of about 40,000 metric tons per year. Thus, if recycled, this amount of slightly contaminated scrap would constitute only an additional 0.1 percent of the annual stream of recycled obsolete steel.

Available data are insufficient to characterize the inventory of concrete SRSM from the DOE complex. One DOE study (DOE, 1996) estimates that about 3.1 million cubic meters (~3.7 million metric tons) of rubble and debris will result from all decontamination and decommissioning operations through 2050. (Together with the estimate of steel SRSM given above, this data implies a mass ratio of concrete to metal of 3.7 to 1—an aggregate number that could vary widely by individual site and type of facility.) Another DOE study (DOE, 1999) has estimated the DOE concrete volume would be over 10 million cubic meters (greater than 12 million metric tons). These two estimates illustrate the kind of uncertainty that exists in the amount of potentially contaminated concrete present in the vast DOE complex.

Much of the concrete will probably be used as on-site fill material, after in situ removal of isolated areas of contamination with an appropriate decontamination technology. As shown in Table 3-6, the quantity of radioactively contaminated soil that may arise during cleanup efforts at DOE facilities could be as large as 76 million cubic meters.

DoD Facilities

Many DoD facilities are licensed by the USNRC, including hospitals, laboratories, proving grounds, some nuclear reactors, weapons facilities, and missile launch sites. The DoD holds approximately 600 licenses and/or radioactive materials permits, of which three-quarters are for sealed sources (and therefore generate no radioactive waste). Most of these licenses cover a spectrum of operations similar to those found in the civilian world. As noted, the USNRC does not license naval reactors and associated propulsion units. Overall, about 115,000 cubic feet of LLRW is generated annually from DoD facilities. Most of this waste (greater than 90 percent) is from cleanup efforts rather than operations.

TABLE 3-6 Sites Containing Radioactively Contaminated Soils

Authority	Location or Type	No. of Sites	Soil Volume (10^3 m^3)
DOE	Fernald	1	2,100
	Hanford	1	23,600
	Idaho	1	720
	Miamisburg	1	110
	Nevada Test Site	1	16,000
	Oak Ridge Reservation	1	133
	Paducah	1	990
	Portsmouth	1	25
	Rocky Flats	1	460
	Savannah River	1	19,000
	Weldon Springs	1	480
	Lawrence Livermore National Laboratory	2	2,212
	Los Alamos National Laboratory	1	9,900
	Sandia National Laboratories	2	221
USNRC or agreement states	Nuclear fuel cycle (active and inactive), including nuclear power plants	199	32
	Byproduct licensees	1,994	60
Other nonfederal	Rare-earth mill sites	17	120

SOURCE: Wolbarst (1999).

EPA-Regulated Superfund Sites

For more than a half century, radioactive materials have been produced and used in weapons production, power generation, and industrial and medical applications. Because these materials were frequently released into the environment, thousands of sites within the United States have become contaminated—some slightly, some heavily. Furthermore, other industrial activities not focused on using radioactive materials have resulted in the concentration of significant amounts of NORM at various sites. As reported by the EPA (63 Federal Register 51982-51888; September 29, 1998), there are about 1,200 sites on the National Priorities List (NPL) of facilities needing cleanup, of which about 150 are federal facilities. According to one estimate, at least 75 sites on the NPL are radioactively contaminated (Wolbarst, 1999). A current estimate by EPA places the number of sites on the NPL having radioactive contamination at approximately 60 (EPA, 2001).

Although DoD and DOE are responsible for the majority of these sites, more than 20 of them did not originate from federal agency activities. Table 3-6 illustrates the approximate inventory of sites containing soils contaminated with radioactivity, their locations, and the estimated volumes of contaminated soil associated with each site.

NORM, NARM, and TENORM

Several types of industrial activity coincidentally enhance the concentration of NORM in waste residues, resulting in the generation of TENORM. The typical radionuclides of concern in TENORM are members of the thorium and uranium decay series. The type of processing performed on natural materials and the time expired since processing determine the equilibrium status of the radionuclides present.

Industries associated with TENORM production may produce radioactively contaminated scrap metals, in addition to TENORM-containing waste residues. These industries include the following:

- Petroleum production,
- Uranium mining,
- Phosphate and phosphate fertilizer production,
- Fossil fuel combustion facilities (power plants),
- Drinking water treatment facilities,
- Metal mining and processing facilities, and
- Geothermal energy production facilities.

Currently, there are no federal statutes explicitly regulating TENORM, although some waste streams fall under the jurisdiction of various EPA regulations or programs. Several agreement states regulate TENORM under their general rules governing possession of radioactive materials, and 11 states have promulgated regulations specifically addressing TENORM. Table 3-7 lists estimates of TENORM wastes generated annually, with associated ranges of uranium, thorium, and radium concentrations. Waste management practices or clearance of

TABLE 3-7 Sources, Quantities, and Concentrations of TENORM

Waste Source	Metric Tons per Year	Concentration[a] (Bq/kg) Uranium	Thorium	Radium
Uranium overburden	3.8×10^4	1.8×10^3	990	920
Phosphate	5.0×10^4	Bkg-3.0×10^3	Bkg-1.8×10^3	400-3.7×10^6
Phosphogypsum	4.8×10^4	Bkg-500	Bkg-500	900-1.7×10^3
Slag	1.5×10^3	800-3.0×10^3	700-1.8×10^3	400-2.1×10^3
Scale	4.5×10^0	—	—	1.1×10^3-3.7×10^6
Phosphate fertilizers	4.8×10^3	740-2.2×10^3	37-180	180-740
Coal ash	6.1×10^4	100-600	30-300	100-1.2×10^3
Fly ash	4.4×10^4	—	—	—
Bottom ash	1.7×10^4	—	—	—
Petroleum production	2.6×10^2	—	—	bkg-3.7×10^6
Scale	2.5×10^1	—	—	bkg-3.7×10^6
Sludge	2.3×10^2	—	—	bkg-3.7×10^3
Petroleum processing	—	—	—	^{210}Pb and ^{210}Po
Refineries	—	—	—	>4.0×10^3
Petrochemicals	—	—	—	>4.0×10^3
Gas plants	—	—	—	^{210}Pb and ^{210}Po
Water treatment	3.0×10^2	—	—	100-1.5×10^6
Sludge	2.6×10^2	—	—	100-1.2×10^3
Resins	4.0×10^1	—	—	300-1.5×10^6
Mineral processing	1.0×10^6	6-1.3×10^5	8-9.0×10^5	<200-1.3×10^5
Rare earths	2.1×10^1	2.6×10^4-1.3×10^5	9.0×10^3-9.0×10^5	1.3×10^4-1.3×10^5
Zr, Hf, Ti, Sn	4.7×10^2	6-3.2×10^3	8-6.6×10^5	300-1.8×10^4
Alumina	2.8×10^3	400-600	500-1.2×10^3	300-500
Cu and Fe	1.0×10^6	<400	<400	<200
Geothermal waste	5.4×10^1	—	—	400-1.6×10^4
Paper mills	—	—	—	>3.7×10^3
Total	2.27×10^6	—	—	

[a]bkg = background radiation level.
SOURCE: USNRC (2001a).

materials from regulatory control depends on both the bulk quantity of the material involved and the concentrations of these key radionuclides in it.

As shown in Table 3-7, the amount of TENORM that could fall under USNRC waste disposal regulations would be about 2.3 million metric tons per year, on a continuing basis.

FINDINGS

Finding 3.1. Licensees may seek to clear about 740,000 metric tons of metallic SRSM that arise from decommissioning the current population of U.S. power reactors during the period 2006 to 2030 (about 30,000 to 42,000 metric tons per year). About 8,500 metric tons per year are expected to arise from decommissioning USNRC-licensed facilities other than power reactors during the same time period. The total quantity of metal from both power reactor and non-power reactor licensees, up to approximately 50,000 metric tons per year, represents about 0.1 percent of the total obsolete steel scrap that might be recycled during that same 25-year period.

Finding 3.2. If most of the licensees of currently operating reactors obtain 20-year license extensions, relatively little SRSM will arise from power plant decommissioning during the 2006-2030 period.

Finding 3.3. Because of the difficulty of determining the quantities and levels of contamination that have penetrated into the concrete, concrete SRSM is generally considered to be volume contaminated. Concrete SRSM constitutes more than 90 percent of the total SRSM arising from decommissioning the population of U.S. power reactors.

Finding 3.4. About 1 million metric tons of metallic SRSM and anywhere from about 3.7 million metric tons to greater than 12 million metric tons of concrete SRSM are projected to arise from cleanup and decommissioning of DOE facilities during the coming 25 years. This quantity of metallic SRSM is comparable in magnitude to the quantity of metallic SRSM estimated to arise from decommissioning the population of U.S. power reactors and corresponds to only an additional 0.1 percent of the total obsolete steel scrap recycled in the United States during the same 25-year period.

Finding 3.5. TENORM is generated in the United States at an annual rate of about 2.3 million metric tons per year. The quantity of TENORM SRSM predicted to arise over the coming 25-year period is nearly 16 times larger than the quantity of SRSM estimated to arise from decommissioning the population of U.S. power reactors.

4

Pathways and Estimated Costs for Disposition of Slightly Radioactive Material

For this discussion, the study committee has assumed that the following three possibilities are available for disposition of the inventories of suspected low-activity radioactive and/or slightly radioactive solid material (SRSM) arising from operating and decommissioning nuclear facilities:

- No release (i.e., disposal to a licensed low-level radioactive waste [LLRW] disposal facility);
- Conditional clearance (release for controlled reuse or disposal in a municipal or hazardous waste landfill); and
- Clearance (unrestricted reuse, recycle, or disposal).

Figure 4-1 illustrates the general decision pathway for disposition under these three possibilities. Under a no-release scenario, all of the materials are sent directly to LLRW disposal. All other disposition scenarios begin with an initial sorting of materials into two streams: cleared materials and materials needing further scrutiny. The materials not cleared are then divided into streams to undergo treatment or not. The uncleared-not treated stream is sorted into a stream for LLRW disposal and a stream of conditionally cleared materials. The post-treatment stream is sorted into three streams: LLRW disposal, conditionally cleared material, and cleared material.

Conditionally cleared material may be released for controlled reuse or disposal in a Resource Conservation and Recovery Act (RCRA) Subtitle D (or, less

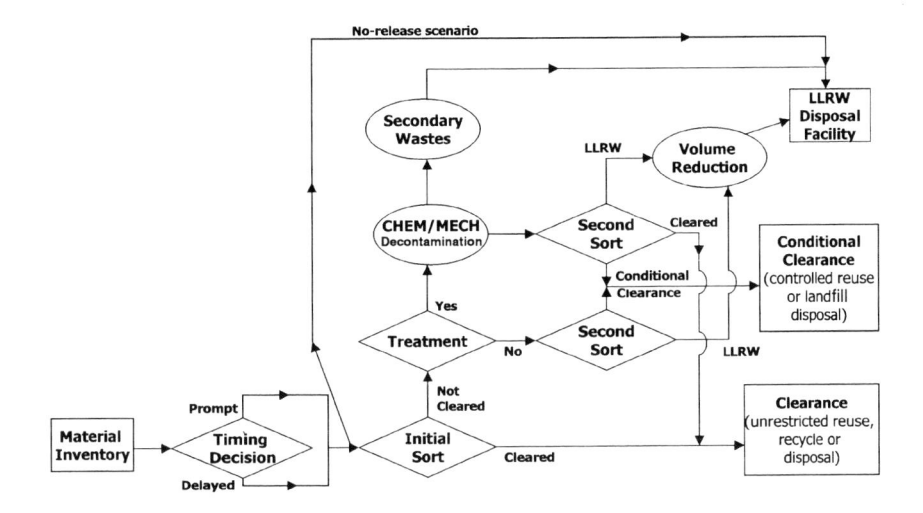

FIGURE 4-1 Decision points and disposition pathways.

frequently, Subtitle C) landfill.[1] Material cleared for disposal should be managed according to its nonradiological properties. In the remainder of this chapter, the committee discusses these pathways and the various decision points in the disposition system represented by Figure 4-1. The discussion includes estimates of the costs for disposing of metallic and concrete SRSM from the population of licensed power reactors in the United States via these three possibilities. Decontamination, segmentation, and transport costs are not included in the costs estimated in this report for disposition.

DISPOSITION SYSTEM DECISIONS

Many nuclear facilities today use waste brokers (firms licensed to receive, process, package, and transport suspected radioactive materials) to handle selected materials arising from their facility operations or decommissioning activities. Thus, an initial decision the waste generator makes is whether to handle its

---

[1] In addition to municipal solid waste landfills (MSWLF), which must meet minimum national criteria set forth by EPA at 40 CFR Part 258, two other types of Subtitle D landfills are commonly used: construction and demolition landfills, and industrial solid waste landfills. The latter two types of landfills are subject to state regulation with respect to liners, leachate collection, etc., requirements which can vary state to state; there were approximately 3,000 such facilities in the United States in the mid-1990s. There is frequent overlap between the types of waste received at the different types of facilities, e.g., regulations allow MSWLFs to receive industrial nonhazardous waste, and in some states construction and demolition waste are disposed of in MSWLFs.

waste disposal activities in-house or to contract with a waste broker. In either case, the decision points and pathways shown in Figure 4-1 remain the same.

The next decision is whether the SRSM could be stored for a sufficient time prior to disposal to allow decay of radionuclides to levels that might meet adopted clearance standards. For this purpose, materials could be stored at either the waste generator's site, a licensed storage facility, or a waste broker's site. Many factors can influence this decision, including the following:

- Will the radioactive isotopes present decay rapidly enough that a reasonably short storage period is possible?
- Is suitable storage capability available either on-site or at a licensed waste broker's site?
- Does the waste owner have the long-term financial stability to ensure safe and proper storage of the radioactive materials and future disposition of the residual material at the end of the storage period?
- Are the avoided immediate disposal costs and the projected future disposal costs and disposal capacities sufficiently well known to justify the risks of a longer-term financial commitment?
- Are the surrounding communities amenable to the long-term storage of these materials?

Some waste generators (particularly hospitals) already use a storage approach for wastes that contain short-lived radionuclides, such as those used in nuclear medicine for treatment or diagnosis. Generally, storage for less than a year is sufficient to permit disposal of these types of wastes subject only to other characteristics that might dictate disposal at hazardous waste (Subtitle C) or municipal waste (Subtitle D) landfills under existing guidance (i.e., Regulatory Guide 1.86). In these circumstances, storage is less costly than the expenses associated with packaging, transport, and disposal at an LLRW facility. In some locations, access to an LLRW disposal facility may be restricted by the compacts and the Low-Level Radioactive Waste Policy Amendments Act of 1985 (LLWPAA). Storage for decay may be the only choice.

Generators of SRSM containing radioactive species with half-lives in the range of one year or less may find the storage approach appealing. However, if the radioactive species have half-lives longer than a few years, the SRSM generator cannot solve the disposal problem with a storage approach.

In the conceptual framework, the next activity is to sort the waste stream into materials that presently can be cleared subject to the appropriate standards and those that cannot. The cleared material is then released for unrestricted use.

The next decision is whether treatment is available and will be used prior to disposition. The SRSM is sorted into two streams, one amenable to treatment and one for which treatment would not be beneficial. Materials to be treated are decontaminated using various chemical or mechanical methods to remove radioactive contaminants from their surfaces. The SRSM that has been subject to treatment is then sorted into cleared, conditionally cleared, and LLRW streams (i.e., no release). The untreated materials are sorted into conditionally cleared and LLRW streams. The two streams of conditionally cleared materials can then be released for controlled reuse or for disposal in a Subtitle C or Subtitle D landfill. The LLRW materials may be reduced in volume before being delivered to an LLRW disposal facility. A secondary radioactive waste stream generated from the chemical or mechanical decontamination activities will also require disposal at an LLRW disposal facility.

RELATIVE COSTS FOR DISPOSITION ALTERNATIVES

Determining the costs for the pathways in this disposition system can be difficult, but some useful data are available. Components of disposal prices at Barnwell and U.S. Ecology are part of the public record. The disposal costs for special items such as reactor pressure vessels or steam generators are often negotiated privately between the waste owner and the disposal facility. In addition, many waste generators now use waste brokers to process and dispose of their wastes. These costs are based on negotiated contracts, which are generally not public record and are therefore not readily available. Many factors affect costs, and the committee was not able to make a detailed analysis of all these factors nor did it find that the U.S. Nuclear Regulatory Commission (USNRC) had prepared a detailed economic analysis. Factors affecting costs include volume, physical and chemical characteristics of the material, taxes and fees imposed by the various regulatory entities, and past relationship of the generator and disposal facility.

The disposal cost for LLRW from decommissioning can constitute a major share of the total cost of decommissioning a nuclear power plant (Konzek et al. 1995; Smith et al. 1996). The USNRC must ensure that utility owners deposit adequate monies into the decommissioning funds to cover the cost of decommissioning their nuclear power plants. Therefore, for the past decade the USNRC has issued a periodic report on LLRW disposal costs, *Report on Waste Burial Charges*, NUREG-1307. The latest revision of NUREG-1307 (USNRC, 2000b) lists the published year-2000 charge rates for LLRW disposal at licensed commercial disposal sites in Richland, Washington ("US Ecology"), and Barnwell, South Carolina ("Barnwell"). It also contains information on the escalation of LLRW disposal costs over recent years and a set of generic rates typically being charged by waste brokers for disposition of contaminated concrete rubble and contaminated metals. These generic rates come from a survey of licensed waste brokers. Thus, some data are available for use in estimating the disposition costs for contaminated materials. NUREG-1307 does not include data for Envirocare of Utah, which is not subject to the limitations of the LLWPAA and was designed specifically to receive high-volume, low-activity waste.

Disposal of commercially generated LLRW and SRSM, as defined in this report, is geographically controlled by the provisions of the LLWPAA. The LLWPAA established the framework for the creation of interstate compacts and granted the compacts the authority to exclude the importation of wastes from outside each compact. At the present time, three disposal facilities are operating in the United States and additional facilities are not likely to be developed in the near future. The US Ecology disposal facility on the Hanford Reservation in Washington takes LLRW and some technologically enhanced naturally occurring radioactive material (TENORM) from states in the Northwest Interstate Compact region (Washington, Oregon, Idaho, Utah, Montana, Wyoming, Alaska, and Hawaii) and, by agreement, the Rocky Mountain Compact region (Colorado, Nevada, and New Mexico). The Envirocare facility, located in Clive, Utah, takes some LLRW and SRSM from all over the country but, out of deference to the Northwest Compact, takes limited wastes from that region. The Chem Nuclear facility in Barnwell, South Carolina, currently takes LLRW from all other states, except North Carolina, although waste receipts at Barnwell will be further limited in the future. The South Carolina state Budget and Control Board has reported, "As you are probably aware, a South Carolina state law passed last year limits the annual volume of waste that can be accepted at the Barnwell site through our fiscal year 2008, which ends June 30, 2008. After that date, the site can only accept waste generated within the Atlantic Compact region. For the current fiscal year, July 1, 2001, through June 30, 2002, the site can accept 80,000 cubic feet, which is a 35 percent reduction from the volume received last fiscal year" (Newberry, 2001).

The following discussion of the estimated costs of disposal is provided for illustrative purposes and does not purport to represent the actual costs that any particular waste generator may incur. The projected dates for reactor decommissioning are too uncertain, as are the interest and discount rates appropriate to those dates, to permit any meaningful present value analyses. In addition, the cost of disposal of nuclear waste will in the future be subject to factors the committee is not able to foresee or take into account in these estimates. For example, the closing of Barnwell to receipt of waste from outside the Atlantic Compact after June 30, 2008, could have an effect on the prices charged by Envirocare of Utah and US Ecology for disposal services. However, the possibility cannot be ruled out that other compacts may open competing LLRW disposal facilities pursuant to the LLWPAA of 1985. Such facilities could accept SRSM generated within compact and, at their discretion, from other compacts. Historically, high disposal costs and lack of access to disposal sites have caused licensees to employ volume reduction (e.g., compaction) and other waste management strategies. This was observed, for example, during the closure of Barnwell to certain states during the 1990s (NRC, 2001). Finally, while the committee has considered the probable future market prices for disposal of waste in developing estimates for the costs of various disposition options, other input variables such as the costs of transporta-

TABLE 4-1 Approximate Costs for Disposal of Solid Material as Low-Level Radioactive Waste (dollars)

Site	Average Price per Cubic Meter	Average Price per Kilogram
Chem-Nuclear—Barnwell S.C.	16,800 (metal or concrete)	13.86
US Ecology—Hanford, Wash.	3,120 (metal or concrete)	2.64
Envirocare of Utah—Clive, Utah[a]	388 (concrete)	0.33

NOTES: The table does not include the cost of decontamination, waste processing, transportation, and handling. Taxes and government charges are included. Nominal waste density is 1,200 kg/m^3.

[a]Envirocare does not publish its rates. The committee was able to verify one set of rates for one customer for 11(e)(2) materials only and cannot state whether this rate is representative for disposal of SRSM in general.

tion, treatment, fees and tariffs, and so forth, have not been included in these estimates. The committee recognizes that the costs of treatment, transportation and handling fees can be substantial; however, since these costs are expected to be case dependent, it was decided not to include them in developing generic cost estimates for disposal.

To estimate the costs of LLRW disposal of metal at the US Ecology and Barnwell disposal facilities, the study committee applied the average costs for disposal at those sites to the inventory of net metallic SRSM (excluding concrete) for the population of U.S. power reactors, as shown in Table 3-1. The average cost for disposal of LLRW materials at the US Ecology disposal facility, adjusted to year-2000 dollars, was about $3,120 per cubic meter. The analogous cost for the Barnwell facility was $16,800 per cubic meter. Table 4-1 lists nominal rates for disposal of solid material used in this discussion. Not all licensees are authorized to dispose of materials at all LLRW facilities, due to the regulatory complexities of the waste compact provisions of the LLWPAA. Envirocare of Utah can accept certain types of high-volume low-activity Class A waste under 10 CFR Part 61, naturally occurring radioactive material (NORM), and 11(e)(2) wastes.[2] With respect to bulk scrap metals that would be generated by power reactor decommissioning, Envirocare's waste acceptance guidelines state that the facility can accept "bulk oversized debris in the form of large pieces of metal, boulders, equipment, etc." (Envirocare, 2001, p. 20). However, Envirocare does not publish prices for disposal of wastes, including large pieces of metal, at its facility. The committee has thus not made estimates of disposal of metals at Envirocare. The cost for disposal as LLRW wastes of net metallic SRSM from

---

[2]Charles Judd, President, Envirocare of Utah, presentation to the committee, March 26, 2001.

the total population of power reactors could range from about $1.6 billion for US Ecology disposal to about $8.8 billion for Barnwell disposal.

The committee estimated the costs for disposal of the same net metallic SRSM at a landfill. The unit costs range from about $30 per metric ton at a municipal waste (Subtitle D) landfill to about $110 per metric ton at a hazardous waste (Subtitle C) landfill. Based on the estimate in Table 3-2 for the weight of metallic SRSM from the population of power reactors, the cost for disposal as conditionally cleared metals would be about $22 million in a Subtitle D landfill or about $81 million in a Subtitle C landfill. The possible income (or cost) associated with clearance of the net metallic SRSM could range from an income of about $22 million (assuming a scrap recycle value of about $30 per ton) to a cost of about $22 million (assuming Subtitle D landfill disposal).

Similar cost estimates arise from consideration of disposition of the concrete SRSM from the population of U.S. power reactors. Envirocare can accept concrete debris for disposal, provided it is Class A waste under 10 CFR Part 61 (Envirocare, 2001, p. 19; see also footnote 2). Envirocare does not publically advertise disposal rates and negotiates disposal rates on a case-by-case basis (see p. 484, *Envirocare of Utah, Inc. v. U.S.*, 44 Fed.Cl. 474 (Fed.Cl., Jun 11, 1999) (NO. 99-76C)). In the absence of direct information, the committee has therefore estimated costs for disposing of concrete from power reactors by using the publicly available contract rate for debris (including concrete) used under contract with the U.S. Army Corps of Engineers (USACE) for disposal of 11(e)(2) wastes at Envirocare of $296.8 per cubic yard ($388 per cubic meter) (USACE, 1998). (The previous year, the contract rate for debris, which includes concrete debris, for the USACE was $427.5 per cubic yard—$559 per cubic meter—illustrating the case-by-case variability in the price of disposing of such wastes at Envirocare.) Disposal of all concrete rubble from U.S. power reactors at Envirocare would cost approximately $2.9 billion. Using the US Ecology and Barnwell disposal charge rates given previously, disposal costs for this concrete as LLRW would range from about $2.9 billion (Envirocare), as noted, to $23 billion (US Ecology), to $123 billion (Barnwell), if all of the concrete is disposed in one site. The text and Table 4-1 show a large difference in disposal costs at the three operating sites. Barnwell and US Ecology are regional disposal facilities under the LLWPAA and, as such, are subject to regional and state surcharges, taxes, and some rate regulation. Envirocare is not a regional disposal facility and is not similarly regulated. The committee cannot explain the differences in rates, nor does the committee know whether the quoted rate for 11(e)(2) disposal at Envirocare is representative of rates for other materials, volumes, or generators. Detailed analysis of the components of disposal costs (e.g., surcharges) is beyond the scope of the committee's task.

Disposal costs for this concrete as conditionally cleared material in a Subtitle D or Subtitle C landfill would range from $265 million to $975 million, depending on the type of landfill utilized. Clearance of this concrete for use in roadway

TABLE 4-2 Estimated Costs for Alternative Dispositions of Slightly Radioactive Solid Material[a] (billion dollars)

Disposal Location	SRSM Metals	SRSM Concrete
U.S. Ecology—Richland, Wash.	1.6	23
Chem-Nuclear—Barnwell, S.C.	8.8	123
Envirocare of Utah—Clive, Utah	Not calculated	2.9
Subtitle C landfill (generic)	0.081	0.98
Subtitle D landfill (generic)	0.022	0.27

[a]Values represent disposal of all material at a given disposal site, and do not reflect any credits that might arise from recycle or reuse of this material.

foundations or other similar unrestricted applications would reduce that portion of the disposition costs associated with disposal to nearly zero. Costs for these and other disposal options for concrete and metal are summarized in Table 4-2.

FINDING

Finding 4.1. Disposal of all slightly radioactive solid materials arising from decommissioning the population of U.S. power reactors into low-level radioactive waste disposal sites would be expensive (about $4.5 billion to $11.7 billion) at current disposal charge rates. Disposal in Subtitle D or Subtitle C landfills would be cheaper ($0.3 billion to $1 billion, respectively). Clearance of all of this material could reduce disposal costs to nearly zero (assumes 100 percent reuse or recycle) or might even result in some income (~$20 million) arising from the sale of scrap materials for recycle or reuse. Decontamination, segmentation, and transport costs are not included in the costs estimated in this report for disposition.

5

Review of Methodology for Dose Analysis

In the United States and internationally, there have been several attempts to provide technical guidance concerning the doses that might be associated with various clearance policies for slightly radioactive solid material (SRSM). As part of its charge, the study committee has reviewed the relevant public reports, as well as various commentaries and critiques of those reports. In addition, the committee met with knowledgeable experts involved in preparing the reports, to clarify specific issues, particularly the reasons why dose factors differ between reports.

Because one of the reports, the draft report NUREG-1640, *Radiological Assessments for Clearance of Equipment and Materials from Nuclear Facilities* (USNRC, 1998b), was prepared for the U.S. Nuclear Regulatory Commission, the committee gave it particular attention. The committee has been able to delve sufficiently deeply into the report to form an overall judgment about its usefulness and to make recommendations for next steps.

Based on its review of technical documents from around the world, the committee has drawn a number of conclusions on technical issues. These findings are collected at the end of this chapter. The body of the chapter supports the findings.

Most of the technical material in this field falls under the rubric of risk assessment, which means it inherits both the strengths and the limitations of this discipline. In particular, although risk estimates can provide useful guidance, they do not substitute for policy decisions on what risks are acceptable. Furthermore, "although the conduct of a risk assessment involves research of a kind, it is primarily a process of gathering and evaluating extant data and imposing science-policy choices" (NRC, 1994).

One of the science policy choices to be imposed involves setting boundaries on the scope of the analysis. In this case, the boundaries involve using radiation dose as a surrogate for health impacts and ignoring other consequences considered to be of lesser significance, such as psychological impacts. When it comes to assigning risk to dose, analysts generally accept standard estimates of dose-risk coefficients established by scientific bodies such as the Committee on Biological Effects of Ionizing Radiation (BEIR) of the National Research Council and the United Nations Scientific Committee on the Effects of Atomic Radiation (NRC, 1990; UNSCEAR, 1988). Nevertheless, the technical reports in this field can assist the USNRC and interested parties in making policy judgments about the clearance of SRSM, as long as the following three conditions are met: (1) the boundaries of the relevant risk assessments must be kept in mind; (2) policy decisions about acceptable risk must be separated from technical issues; and (3) the major limitations of the technical reports, as identified in this chapter, must be addressed. The flow chart in Figure 5-1 shows points at which technical information can inform decision makers about clearance of SRSM, if a rulemaking process one day advances to a decision point about clearance.

KEY TECHNICAL ASSESSMENTS OF ANNUAL DOSES ASSOCIATED WITH CLEARANCE OF SOLID MATERIALS

A great deal of effort in a number of countries over the last 20 years has gone into developing the numerical coefficients, also called *dose factors*, needed by policy makers to (1) understand the dose commitment implied by various clearance concentrations and (2) convert a primary dose standard into secondary activity standards that can be used by licensees to ensure compliance with the primary standard (see Box 5-1). The major compilations of these dose factors are listed in Table 5-1, along with the scientific bodies that have reviewed the underlying technical analyses. (See Appendix D for a summary of efforts here and abroad on SRSM clearance standards.)

All of the reports in Table 5-1 estimate doses to classes of persons, such as SRSM transport workers, or consumers, and focus on the group that is estimated to have the highest dose under all the scenarios considered (the *critical group*); see Figure 5-2. The principle is that if the most exposed group of individuals is identified correctly and the dose to that group is shown to fall below the primary standard, then the dose to any other member of the general public will fall below the standard as well. Thus, the dose to the critical group (for a unit release) determines the dose factor. Note that the critical group can differ for different radionuclides, which complicates implementation of any clearance standard that relies on dose factors. Also, the critical group, and thus the secondary standard, may change when the allowed clearance categories are restricted, as in condi-

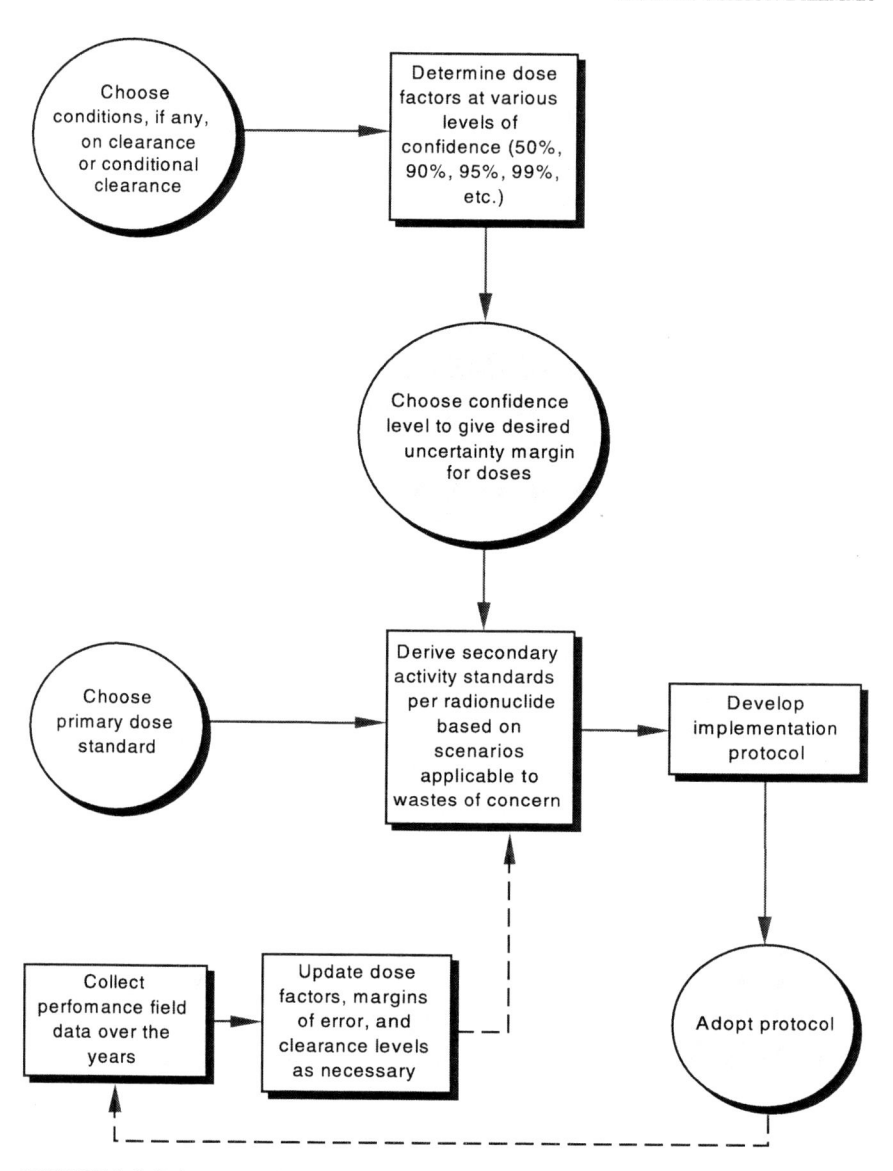

FIGURE 5-1 Points at which technical information and judgments can inform rulemaking decisions related to clearance of slightly radioactive solid material. NOTE: Circles indicate policy decisions. Rectangles indicate technical contributions.

tional clearance. To date, most of the attention to dose factors has assumed that they would be used in setting standards for (unconditional) clearance.

Because primary dose standards for clearance or disposition of solid materials are usually given in dose per year, the dose factors are generally expressed in

> **BOX 5-1**
> **Primary and Secondary Clearance Standards and Dose Factors**
>
> A dose limit for an individual, such as 10 μSv/yr (1 mrem/yr), constitutes a *primary standard* for clearance of a radioactive or slightly radioactive solid material. *Secondary standards* or *derived activity standards* are derived from a primary standard. They apply to a licensee's material and specify the maximum activity in or on a solid material that has been estimated to be clearable while remaining below the primary dose standard. Derived activity standards are set using the results of risk analyses that determine the annual dose of radiation received for a given radionuclide concentration corresponding to a "critical" (most exposed) group of individuals. When developing regulations, analysts construct several groups of scenarios corresponding to different phases of the recycle and reuse of the slightly radioactive solid material during which there is the possibility that persons can be exposed to radiation. These include handling and processing scenarios, storage scenarios, product use scenarios (e.g., if steel is recycled to make a product), transport scenarios, disposal scenarios, and landfill resident scenarios. For a given radionuclide concentration (becquerels per gram), the annual dose to individuals in each of these scenarios (and the subgroupings within them, e.g., transport of scrap metal and transport of slag) is calculated. The critical group in draft NUREG-1640 for steel and cobalt-60, for example, is transport of scrap metal, meaning that the dose received in all other scenarios is lower. The dose factor for cobalt-60 is 250 microsieverts/yr per Bq/g, which is the mean of the distribution reported in NUREG-1640.
>
> The derived activity standard for each radionuclide can be derived from its critical group dose factor by dividing the desired primary dose standard, e.g., 10 microsieverts, by the dose factor. This relationship can be shown by solving for the quantity *x* in the equation below to determine what quantity of Bq/g would cause a dose of 10 microsieverts per year, using the dose factor for cobalt-60 in the critical group just described for steel, as follows:
>
> $$250 \frac{\mu Sr/yr}{Bq/g} \times xBq/g = 10\mu Sv/yr$$
>
> In the case of cobalt-60 in steel, this yields a derived clearance standard of 0.04 Bq/g.

units of dose per year per unit of activity released.[1] Cumulative total doses can be obtained by multiplying the estimated dose by an assumed duration of exposure—for instance, a person's remaining years of life.

---

[1]Presumably, the total dose per year (i.e., the committed dose per year). In some cases, effective dose equivalent is used, which accounts for both the relative biological effectiveness of different types of radioactivity and the differing sensitivity of organs to cancer mortality. In some studies, only effective doses are used, without the weighting by cancer mortality that produces dose equivalents.

TABLE 5-1 Technical Analyses Supporting Numerical Coefficients for Deriving Secondary Activity Standards from Primary Dose Standards

Study	Status	Reviewer	Reference
USNRC			
NUREG-1640[a]	Draft	CNWRA	USNRC, 1998b
NUREG-0518	Draft		USNRC, 1980
EPA			
TSD 97[a]	Draft	NCRP	EPA, 1997a
TSD 99[b]	In progress		
ANSI/HPS			
N13.12-1999	Final		ANSI/HPS, 1999
IAEA			
Safety Practice No. 111-P-1.1	Final		IAEA, 1992
Technical Document 855	Interim		IAEA, 1996
European Commission			
Radiation Protection-89	Final		EC, 1998b
Radiation Protection-114	Final		EC, 2000

NOTE: ANSI/HPS = American National Standards Institute/Health Physics Society; CNWRA = Center for Nuclear Waste Regulatory Analyses; EC = European Commission; EPA = Environmental Protection Agency; IAEA = International Atomic Energy Agency; ICRP = International Commission on Radiological Protection; NCRP = National Council on Radiological Protection and Measurements; TSD = Technical Support Document.

[a]The coefficients given in the USNRC and EPA source documents have built into them, or the opportunity to use, an explicit margin to account for uncertainty. In EPA TSD 97 a margin was built into the dose coefficients. Specifically, the semiquantitative uncertainty analysis described in Chapter 10 showed that, depending on choice of input parameters, normalized doses could be higher by a factor of 5-50 or lower by a factor of 100-500, i.e., they favored more protective levels. The NUREG-1640 draft shows a distribution of dose factors based on Monte Carlo simulations of the aggregate uncertainty resulting from uncertainties in the component estimates. Both the mean values and the 95th percentile given in NUREG-1640 for the dose coefficients lie above the median, 50th percentile value. If either of these properties of the distribution were chosen to define the regulatory dose coefficients, a margin above the best estimate (median) would automatically be included.

[b]The study committee has seen the first EPA report, TSD 97. A new report, TSD 99, was prepared and given very limited distribution, presumably in 1999. Proposals were due to the EPA on April 13, 2001, for final revision of TSD 99, with submittal of the draft to the EPA by May 31, 2001. Both TSD 99 and this new revision will supplant Chapters 1-7 of TSD 97, using ICRP-68 guidance. By this revision the EPA will be reacting to comments from the NCRP review and others. The remaining chapters of TSD 97 after Chapter 7 will apparently stand without revision. (Information on status of TSD 99 and revision efforts was received in a personal communication from Debbie Kopsick, EPA, to Robert Bernero, Board on Radioactive Waste Management, National Research Council, April 11, 2001.)

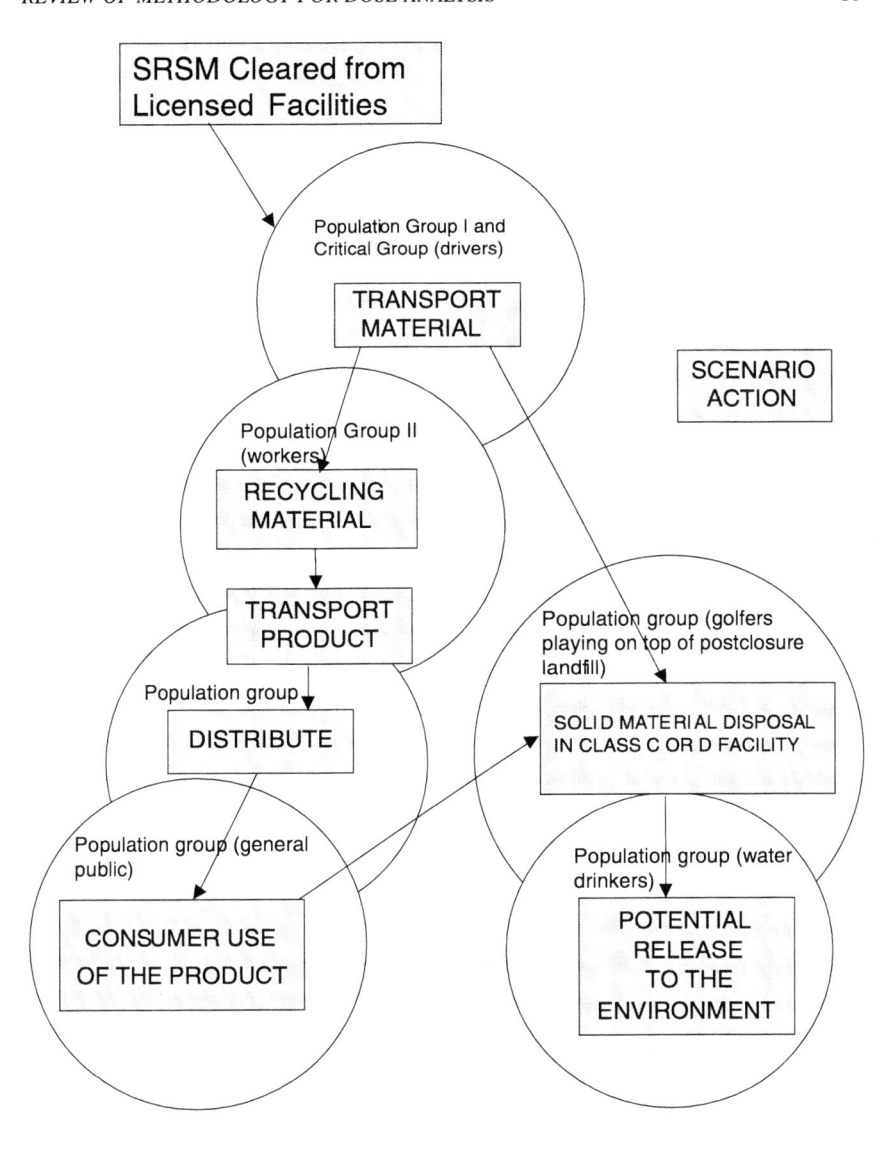

FIGURE 5-2 Illustration of scenario pathways following SRSM clearance and hypothetical affected critical groups.

The committee's assessments of the individual technical sources listed in Table 5-1 are presented in the next five sections. Then the committee compares the methodologies used across these studies, including comments on the usefulness and quality of the dose factors they contain, general limitations that should be corrected, and potential inconsistencies in the dose factors used by different countries. Before concluding with the summary statement of the findings for the chapter, the committee explores in further detail specific issues that should be addressed in subsequent work on the draft NUREG-1640.

USNRC STUDIES

The committee reviewed two technical documents on clearance standards developed for the USNRC. Draft NUREG-1640, which has been mentioned in earlier chapters of this report, is particularly relevant to the new rulemaking on clearance standards for SRSM, which the Commission is contemplating. The second document, NUREG-0518, represents an earlier effort at analysis to support clearance standards for SRSM.

Draft NUREG-1640

Draft NUREG-1640 (USNRC, 1998b) contains estimates of the *total effective dose equivalent* to an average individual in a critical group from direct reuse of equipment, recycling, or disposal of materials, for a wide range of radionuclides that may be present in solid materials from decommissioning of nuclear facilities. The risk assessment methodology is largely state of the art. Critical groups are chosen by assuming a policy of clearance, although information in the appendixes may be sufficient to allow choices of other critical groups to support derivation of dose factors for possible conditional clearance policies. The draft does not discuss implementation issues.

Although NUREG-1640 is a draft for review and comment, it is a sophisticated product and does many things well. The various scenarios considered for clearance of materials with surface or volume contamination are well documented and easy to understand. The major analytical effort is for recycling steel (31 scenarios), with less analysis for recycling copper (23 scenarios), aluminum (17 scenarios), and concrete (7 scenarios). There is an in-depth analysis of current recycling practices and how the inclusion of SRSM would show up as exposure to humans. In addition, the study does a good job of documenting the impact of equipment reuse.

The chemistry, metallurgy, geology, and physics used in the report seem reasonably sound. Considerable information is provided on the dose factors resulting from external exposure, inhalation, and ingestion of radioisotopes from recycled material, waste, and release of effluents to air or water. Most of the critical groups turn out to be workers, not the public at large. The report does not discuss whether this pattern would change for conditional clearance.

The committee found the overall *conceptual* plan of draft NUREG-1640 to be the best of all of the studies that it reviewed. It is closest in spirit to recommendations on risk assessment that have been made by expert bodies, including committees of the National Research Council (NRC, 1994). For instance, the estimates in draft NUREG-1640 are traceable, and a formal uncertainty analysis has been performed for each dose factor.[2] The study presents the mean and the 5th, median, and 95th percentile values for each dose factor, derived from Monte Carlo uncertainty analyses. The authors of draft NUREG-1640 use the range from the 5th to the 95th percentile to define a "90 percent confidence interval" (about the median) (USNRC, 1998b, Tables 4.10 and 4.11).

The result of a Monte Carlo calculation, such as carried out by NUREG-1640, is a distribution of doses for each scenario delivered to the representative member of a critical group for a particular radionuclide. There is no single dose estimate to a critical group, and hence no single dose factor for that critical group. Nevertheless, a decision must be made, if NUREG-1640 is to be used to support clearance or conditional clearance, about which dose factor should be used to assign secondary activity standards. If one takes the median of the distribution, then 50 percent of the dose factors are below and 50 percent above. Choosing the median as the de minimis value for use in clearance or conditional clearance standards, however, would leave the decision maker without the higher degree of assurance that the dose to the critical group is below the ordinary dose standard as when higher percentile-valued dose factors are chosen (e.g., 90th percentile). This additional assurance is above and beyond the conservatism that applies to individuals within population groups that receive less exposure than the critical group.

Table 5-2 averages the uncertainty factors computed for NUREG-1640 across radionuclides. In this table, uncertainty is represented by the geometric standard deviation (GSD), which is appropriate for quantifying the spread in variables with large variations. Except for concrete recycle, the GSDs are small.[3]

---

[2] For the uncertainty analysis, NUREG-1640 works with the individual steps involved in making a dose estimate. The analyst gathers data from the literature on the ranges that individual parameters required for the estimate might take and then propagates the individual uncertainties to the final coefficient using techniques such as Monte Carlo simulation (EPA, 1996). There is a subjective element in choosing the parameter distributions used to fit the literature data, but these choices are one or more steps removed from the final uncertainty estimate for the dose factor. Also, guidelines exist for selecting the functional form for a parameter distribution (Seiler and Alvarez, 1996).

[3] A sampling of papers published in *Health Physics* showed GSDs ranging from 1.7 to 20, with most in the range 2-4. Thus, values of GSD below 2 can be considered small, and values above 4 considered high (Breshears, 1989; Johnston, 1991; Till, 1995; Sheppard, 1997; Bolch, 2001).

TABLE 5-2 NUREG-1640 Uncertainty Factors Averaged Across Radionuclides

	Average Geometric Standard Deviation (GSD)[a]	
	Volume Contamination	Surface Contamination
Steel recycle[b]	1.8	2.0
Concrete recycle[c]	3.0[d]	3.4
Copper recycle[e]	1.5	1.6
Aluminum recycle[f]	1.4	1.7
Reuse of large piece of equipment[g]	NA[h]	1.9

[a]One standard deviation is equal to the product of the median times the GSD. Two standard deviations (~95th percentile limits) equal the square of the GSD. For the table, GSDs were approximated by computing the square root of the ratio of the 95th percentile dose factors to the 50th percentile results, as presented in tables in NUREG-1640 (USNRC, 1998b).
[b]From Tables 4.1, 4.2 (USNRC, 1998b).
[c]From Tables 7.2, 7.3 (USNRC 1998b).
[d]The distribution is bimodal, with one group of radionuclides having a GSD around 1 and another group having a GSD around 6.
[e]Tables 5.5, 5.6 (USNRC, 1998b).
[f]Tables 6.4, 6.5 (USNRC, 1998b).
[g]Tables 3.2, 3.3 (USNRC, 1998b).
[h]NA = not applicable.

Formal uncertainty analysis can be an important tool for building confidence in the use of dose estimates for policy decisions. It addresses the reported tendency of even experts in a field to underestimate uncertainty bands when professional judgment alone is used (Cooke, 1991). This tendency exists even in the physical sciences (Shlyakhter and Valverde, 1995). It is therefore not wise to rely on professional judgments of estimates of overall uncertainty because of the subjective bias found in such estimates. Of the studies listed in Table 5-1, only draft NUREG-1640 includes a formal uncertainty analysis that reduces the amount of professional judgment required in assigning uncertainty bands to dose factors. Excellent discussions of formal uncertainty analysis can be found in other USNRC documents (e.g., USNRC, 1995) and in Morgan and Henrion (1990).

The authors and planners of draft NUREG-1640 are to be commended for developing an excellent approach. The execution of draft NUREG-1640's conceptual plan, however, has been clouded by questions of contractor conflict of interest concerning the recycle option (see discussion in Chapter 2). One question is how the USNRC could have failed to identify the conflict of interest. These questions highlight the need to include the possibility of organizational failure when assessing overall system uncertainty. After the conflict of interest was identified, the matter was investigated by USNRC counsel and the contract in question was terminated. The USNRC engaged another contractor to complete the work on draft NUREG-1640.

Meanwhile, the USNRC asked the Center for Nuclear Waste Regulatory Analyses (CNWRA) to perform an independent technical review of the draft NUREG-1640. The CNWRA, located at the Southwest Research Institute in San Antonio, Texas, is a dedicated contractor providing technical support to the USNRC on waste management matters. The committee has studied the CNWRA review (CNWRA, 2001), which is actually an audit of the mathematics and completeness of scenarios considered in draft NUREG-1640. CNWRA recommended that some additional scenarios be added to the mix considered in draft NUREG-1640 but otherwise found the mathematics to be correct. Although this CNWRA review is comforting and is confirmed by the committee's spot check of some of the scenarios, there has not yet been a similarly thorough review of the choice of parameters and parameter ranges, term by term, for the component estimates in deriving the dose factors. (The choice of parameters and parameter ranges are listed in Appendix B of draft NUREG-1640, Tables B.1, B.2, etc. Although the committee generally confirmed the reasonableness of many of these choices, it was able to review only a sample of the dose factors, given its full set of tasks.)

In addition to any lingering questions about the choice of parameters, whether due to a potential bias or for other reasons, there are a number of other limitations in draft NUREG-1640. These limitations have to be addressed before the document will be fully usable by the USNRC and interested parties in reaching valid conclusions about related SRSM policy issues. The limitations are discussed in the penultimate section of this chapter. One option for the USNRC, faced with any lingering concerns over charges of conflict of interest, is to start all over again. However, it is likely that any new contractor would simply repeat the work in NUREG-1640 as far as it goes and build upon it in the way the committee recommends. Therefore, from a scientific perspective, the committee does not believe it is cost-effective to repeat the work done in draft NUREG-1640.

The committee believes that once the remaining questions about and limitations in draft NUREG-1640 are addressed, either in the final version of the report or in follow-up reports, the USNRC and interested parties will have a sound technical basis for evaluating the health impacts, measurement issues, and implementability of various primary dose standards and the unavoidable uncertainties involved in risk estimates. However, the committee notes that the dose factors developed through the NUREG-1640 process cannot be adopted for use with Department of Energy (DOE) or other SRSM without further analysis. Changes are likely to be needed to some of the dose factors and/or their uncertainties because the quantity and types of DOE SRSM, as well as some potential release scenarios, differ from wastes generated by USNRC-licensed facilities.[4]

NUREG-0518

Prior to NUREG-1640, the USNRC published a risk assessment in 1980 for the release of SRSM. In response to a 1974 amendment to the Atomic Energy Act (AEA), which authorized release of de minimis quantities of special nuclear material if justified, the Atomic Energy Commission (AEC) began developing a de minimis standard for enriched uranium and the attendant fission product technetium. The development side of the AEC (later the Energy Research and Development Administration, ERDA) requested guidance from the regulatory side of the AEC (later the USNRC). The ERDA developed data for the quantities of scrap steel, copper, and nickel that would become available from the 1976-1982 cascade improvements at the gaseous diffusion plants used for enriching uranium. (See also "DOE Facilities" in Chapter 3.) Data were provided on the extent of decontamination that could be achieved by smelting. These data showed that smelting could not be relied upon to reduce the contaminant content to less than 17.5 parts per million (ppm) uranium and 5 ppm technetium.

The USNRC staff prepared and issued NUREG-0518, *Draft Environmental Statement Concerning Proposed Exemption from Licensing Requirements for Smelted Alloys Containing Residual Technetium-99 and Low-Enriched Uranium* (USNRC, 1980). NUREG-0518 contained analyses of the expected scrap metal inventories from the gaseous diffusion plants and of scrap metal from other sources. Using several important assumptions, the study estimated both doses to an individual member and collective doses to the entire group for several critical groups. The most important assumption was that the proposed exemption from regulatory control would apply only to scrap metal ingots coming out of a licensed smelter, thus ensuring a radionuclide content in the scrap of no more than 17.5 ppm uranium and 5 ppm technetium. Conservative assumptions were ap-

---

[4]DOE may have disposition opportunities, and therefore clearance scenarios, that are not available to USNRC licensees. As for dose calculations, the committee notes that uncertainties about migration of transuranics have become important for DOE SRSM, whereas they are far less important for SRSM from USNRC licensees. For example, the transuranic radionuclides in the SRSM stream from USNRC-licensed facilities constitute a relatively minor component of the radioactive contaminants. USNRC analysts therefore do not need to delve too deeply into chemical and biological processes in landfills that might speed up migration of transuranic radionuclides, which are thought to migrate at a slow rate under usual subsurface conditions. By contrast, DOE has a great deal of material potentially contaminated with transuranics at substantially higher concentrations than occur in nuclear power plant wastes. An analysis by DOE to support conditional clearance standards for DOE SRSM may have to consider in some detail the chemical or biochemical processes in Subtitle C or other landfills.

plied to scenarios for possible uses of the ingots. For example, all of the steel scrap released was assumed to be used in a continuous 80-day run at the exempt steel plant and made into products of the reference content. Steel plate, iron tonic, and even a production run of 9 million cast iron frying pans, were considered as possible products from the steel. Jewelry, coins, and prostheses were considered as possible products from the other metals.

The estimated doses listed in NUREG-0518 include a 10 mrem/yr wholebody external dose for one exposed group (workers spending 1,000 hours per year in a steel vault), a 2 mrem total-body dose commitment for another group (1 year of iron tonic ingestion), and a 20 rem contact bone dose for a third group (prosthesis pins implanted for 50 years). The collective dose for the worst-case scenario was estimated to be 80 person-rem.

NUREG-0518 does not contain any uncertainty analysis as such. Instead, it invokes conservative bounding conditions to make the point estimates of dose usable for regulatory purposes.

In NUREG-0518 the USNRC staff concluded that the proposed exemption, as qualified, was acceptable for consideration by the Commission for amendment of its regulations. There was substantial negative public reaction to NUREG-0518, and the proposed exemption process was suspended (51 Federal Register 8842; March 14, 1986).

ENVIRONMENTAL PROTECTION AGENCY DOCUMENTS ON DOSE FACTORS

The Environmental Protection Agency (EPA) Technical Support Document (TSD) *Evaluation of the Potential for Recycling of Scrap Metals from Nuclear Facilities* ("TSD 97") contains numerous tables of background information on the sources and inventories of radioactively contaminated metal scrap from various government and commercial sources (EPA, 1997a). The document develops various normalized individual doses, collective doses, and collective risks, normalized to curie-per-gram concentrations in the scrap metal streams. It also contains valuable information, compiled in an insightful way, about detection limits for contamination as a function of various parameters and about various scrap metal processes, including how radionuclides partition in these processes. This is all useful information. The methodology employed and the handling of uncertainties helped the study committee understand the relevant issues.

TSD 97 also contains useful discussions about possible pathways from contaminated metal (sources) to humans (receptors). These pathways are sorted into a few important pathways and a much larger number that were judged to be less important. The basis for the sorting is explained well.

Another useful element is the discussion of an estimated "timetable," covering the next few decades, indicating when the waste streams might become available for potential commercial recycle (or other disposition alternatives).

This discussion, although inexact in detail because of some assumptions that cannot be verified, succeeds in putting the issues in context.

To assess the uncertainty associated with doses to the critical group, TSD 97 performed a semiquantitative uncertainty analysis that "evaluated the uncertainty/variability in the dose evaluation results due to uncertainty/variability in the calculational parameters and assumptions."[5] Although not a formal uncertainty analysis, the analysts used their inspection of these results and professional judgment to conclude that the dose factors they calculate represent a 90th percentile. (That is, in 90 percent of cases, use of the calculated dose factor will result in a dose to a member of the critical group that is at or below the primary dose standard.[6])

The National Council on Radiation Protection and Measurements (NCRP, 1998) has produced a detailed critique of TSD 97. Among its major findings and recommendations are the following (NCRP, 1998, pp. 9, 11):

1. The NCRP task group concluded that, "as it now stands, [TSD 97] overemphasizes the evaluation of a limited number of scenarios with data that are incomplete and/or unsupported."
2. The NCRP task group recommended "the use of a probabilistic risk assessment model, such as the Monte Carlo method (as recommended by [the EPA's] established policy relative to the conduct of [probabilistic risk assessments]), for analyzing the potential uncertainties and for identifying areas for improvements in the input data."
3. The NCRP task group recommended that the EPA evaluate the feasibility for implementation, stating, "Standard development cannot be devoid of information regarding implementation."

These comments from the NCRP task group are apparently being taken into account by the EPA as it works on a revision of TSD 97 (EPA, in progress). The committee has not seen the revision, which was still in progress when the various technical documents on dose factors were being reviewed for this report.

AMERICAN NATIONAL STANDARDS INSTITUTE AND HEALTH PHYSICS SOCIETY STANDARD N13.12-1999

The Health Physics Society (HPS) Standards Working Group developed this standard.[7] The document defines primary (dose) and secondary screening (activity level) criteria (ANSI/HPS, 1999). The primary dose standard is

---

[5] TSD 97 (EPA, 1997a, p ES-8, see also, Ch. 10, p. 12).
[6] TSD 97 (EPA, 1997a, Ch. 3, p 3).
[7] The standard was consensus balloted and approved by the ANSI-accredited HPS N13 Committee on October 19, 1998. It was approved by ANSI, Inc., on August 31, 1999.

10 µSv/yr (1 mrem/yr), which is consistent with international values. The document tabulates derived screening levels, above background, for the clearance of SRSM or items containing surface or volume activity concentrations of radioactive materials. These screening levels are derived by applying dose factors to the primary dose standard.

The ANSI/HPS document contains a great deal of useful information on uncertainties in dose factors. Furthermore, the working group took on the difficult task of developing an implementation protocol, which specifies areas over which measurement averages should be taken. It also groups radionuclides based on similarity of dose factors and assigns group-level screening levels ranging from 0.1 to 100 Bq/cm^2 or Bq/g, depending on the group considered. The dose factors chosen are quite similar to the International Atomic Energy Agency (IAEA) values.

To derive dose factors, the working group reviewed a range of dose estimates produced by different analysts for different activities, such as landfill disposal and steel recycling. It also used reports that examined exposures for different forms of contamination (either volume or surface contamination). In contrast to other reports the committee has reviewed, the working group did not use the range of dose estimates across categories to define a critical group in a documented manner. As a result, the method for deriving the screening levels is not traceable by independent reviewers.[8] Although the ANSI/HPS working group was composed of analysts of great skill and experience, only a traceable approach could be judged and ranked by the committee.

INTERNATIONAL ATOMIC ENERGY AGENCY DOCUMENTS

The committee reviewed two documents developed by the International Atomic Energy Agency: Safety Practice No. 111-P-1.1, *Application of Exemption Principles to the Recycle and Reuse of Materials from Nuclear Facilities* (IAEA, 1992), and a more recent interim document, IAEA-TECDOC-855, *Clear-*

---

[8]Based on a discussion with a working group member, it appears that the working group used professional judgment to discount or reduce dose values from scenarios if the group believed the value to be unreasonably conservative. It then picked the highest remaining value to use in setting screening levels (personal communications from William Kennedy, HPS Standards Working Group, to Jan Beyea, committee member, April 20, 2001). Had the working group included a table in the standard with the discounted factors, the methodology would have been traceable.

As part of the working group's analysis, it concluded that dose factors appeared to be similar for surface and volume contamination, when units were expressed in becquerel per gram or becquerel per centimeter squared. (IAEA, 1996, came to a similar conclusion.) Consequently, the group chose the same "derived screening levels" to apply to both surface and volume contamination in the implementation protocol. Again, no summary of the values from which the group drew its conclusions was included in the report, making its analysis untraceable. The study committee recognizes that a volunteer group, such as the HPS Standards Working Group, can include only a limited amount of detail in its reports.

ance Levels for Radionuclides in Solid Materials: Applications of Exemption Principles (IAEA, 1996). Comments on each document are presented below.

Safety Practice No. 111-P-1.1

In Safety Series No. 89, *Principles for the Exemption of Radiation Sources and Practices from Regulatory Control*, the IAEA established the principles that underlie its technical estimates of dose factors (IAEA, 1988). The USNRC has produced no similar generic document. The IAEA dose factors are contained in Safety Practice No. 111-P-1.1, *Application of Exemption Principles to the Recycle and Reuse of Materials from Nuclear Facilities* (IAEA, 1992). Two of the IAEA recommendations from these documents may differ from the concept of clearance of SRSM under discussion in the United States:

1. "The dose to the individual of the critical groups(s) and the dose to the whole population exposed by the practice should not be significantly affected by other similar (or identical) practices (e.g., several waste disposal sites in the same region)" (IAEA, 1988, p. 6).
2. "The formulation of an exemption should not allow the circumvention of controls that would otherwise be applicable, by such means as deliberate dilution of material or fractionation of the practice" (IAEA, 1992, p. 4).

The technical calculations for Safety Practice No. 111-P-1.1 were completed in 1993. The authors considered recycle of steel, aluminum, and concrete. They also analyzed reuse of surface-contaminated rooms in buildings and reuse of tools and equipment. The report contains no uncertainty analysis. Instead, a conservative approach was taken to deterministic calculations. Parameters were assigned values from the upper end of their observed or expected ranges. This approach produces results that "are likely to overpredict doses which will be received in practice (if they are received); however, it is difficult to say by how much they are higher than the 'real' values" (IAEA, 1992, p. 49).

A Monte Carlo analysis was carried out for ^{60}Co in asphalt, which confirmed that the base case estimate produced an overestimate of exposure (IAEA, 1992, pp. 104-105). In addition, a limited sensitivity analysis was undertaken for steel recycling to study the effects of three basic assumptions on the partition, dilution, and quantity of contaminated steel (IAEA, 1992, p. 49).

A limitation in the report is the use of values for some parameters without citing sources,[9] which makes it difficult for independent reviewers to trace the analysis.

---

[9]See, for example, Appendix II, p. 97, of IAEA (1992), where values for a resuspension factor, the fraction of surface contamination available for resuspension, the rate of secondary ingestion of removable surface contamination, and the transfer factor for secondary ingestion are given without citation.

Interim Report IAEA-TECDOC-855

In 1996 the IAEA prepared an interim report *Clearance Levels for Radionuclides in Solid Materials: Application of Exemption Principles*, in which it reviewed a set of studies, including its own, to pick a set of dose factors to use in deriving secondary activity standards for clearance (IAEA, 1996). The secondary standards are derived by dividing the primary standard recommended by the IAEA (10 mSv/yr) by the dose factor that the authors decided on for each radionuclide. A similar approach was later used by the HPS Standards Working Group to prepare the ANSI/HPS clearance standard. However, unlike ANSI/HPS, the IAEA study includes the steps the authors took to discount various studies, so the work is traceable. To simplify implementation, the authors grouped radioncuclides with similar clearance levels by rounding values. No uncertainty analysis is presented in the report.

EUROPEAN COMMISSION DOCUMENTS

The European Commission (EC) has produced a number of technical and policy documents that deal with clearance issues. The two main technical reports are EC-RP-89, *Recommended Radiological Protection Criteria for the Recycling of Metals from the Dismantling of Nuclear Facilities* (EC, 1998b), and EC-RP-114, *Definition of Clearance Levels for the Release of Radioactively Contaminated Buildings and Building Rubble* (EC, 2000). These reports address metals recycling, equipment and building reuse, and building demolition.

For buildings and building rubble, the analysts used a few scenarios that are assumed to be representative of the many others that have been studied by other analysts. An analysis assuming homogeneous volume contamination produced "nuclide specific clearance levels" (i.e., secondary standards) that were prohibitively restrictive for large buildings, so the authors took into account the likelihood of inhomogeneous contamination and other factors to reduce the clearance levels by a factor of 10 (EC, 2000). An explicit assumption in the EC analyses, which is built into the EC recommendations, is that it is forbidden to mix highly contaminated surfaces or rubble with the uncontaminated bulk of the structure.

Apparently, no uncertainty analysis was carried out.[10] Presumably the underlying doses were calculated with a tendency to choose individual parameters that produced an overestimate in dose, but no statement to that effect is included in the reports. However, the study committee has not reviewed the full consultant's report for EC-RP-114, only what is included in the report itself.

In deriving nuclide specific clearance levels, the EC reports use a collective dose standard of 1 person-sievert (person-Sv) per year and a derived dose stan-

---

[10] The authors did consider what they called "pessimistic" assessments in developing dose factors and clearance values.

dard for individuals of either 10 µSv/yr (1 mrem/yr) or a skin dose of 50 mSv/yr (5 rem/yr) (EC, 1998b, p. 4). If the collective dose exceeds the 1 person-Sv/yr standard, a decision must be made on whether the activity has been optimally reduced, (i.e., is as low as reasonably achievable [ALARA]).

This approach suggests a refinement that the USNRC should consider as it deliberates over clearance standards. Suppose that the variations in contaminant levels of a material were so large that the highest values surveyed exceeded the allowed dose to a member of the public, even though the average value was at or below the USNRC clearance standard. It might be desirable to require reduction of the activity level to the point that the dose standard was not exceeded by the highest survey reading.

COMPARISON OF CLEARANCE STUDIES

Table 5-3 compares specific features of the general methodologies used in the studies reviewed by the committee. Not surprisingly, the studies do not always agree on the numerical values for best estimate. To capture the rough magnitude of these differences, Table 5-4 shows the average of the ratios of the NUREG-1640 dose factors to the dose factors presented in other studies. Note that Table 5-4 uses the *mean* NUREG-1640 dose factor coefficient, which lies somewhere between the 50th and 95th percentile values for the dose factor, depending on the radionuclide.

On average, the dose factors for metals in the draft NUREG-1640 and the EPA study are in relatively good agreement. Using the computation explained in Table 5-4, the NUREG-1640 values are lower but on average are within a factor of two of the EPA values. With respect to the dose factors selected in the IAEA and EC reports however, the NUREG-1640 values are on average about 5 to 14 times higher and hence would allow less activity to be released on average given the same primary dose standard. For particular radionuclides and particular critical groups, the disagreement between the U.S. dose factors (NUREG-1640 or EPA TSD 97) and those from the EC studies can be much greater than a factor of 10. For instance, the draft NUREG-1640 dose factor for ^{60}Co is 200 times more restrictive than the EC value for clearing surface-contaminated metals (USNRC, 1998b, Table 2.5).

One reason that dose factors computed for different studies vary is that different simplifying approximations are used. Another reason is that different critical groups and different exposure scenarios for those groups are selected to model doses. In some cases, heterogeneity of contamination was assumed, from which one could derive a lower dose in a given exposure scenario than if uniform contamination were assumed, and therefore increase the activity level allowed for clearance. For example, the EC studies estimate that "the mass specific activity averaged over the total quantity of building rubble (10^5 metric tons) will be around one order of magnitude less than the clearance level" (EC, 2000).

Similar assumptions, which have the effect of reducing the dose factor (and therefore allowing a higher secondary standard [see Box 5-1]) have not been introduced into the analyses from which either the EPA or the USNRC dose factors were estimated. These and other differences in methodology explain some of the difference between the European dose factors and those from the EPA or USNRC studies. Finally, different degrees of conservatism may have been built into the estimates. Large differences do not necessarily imply that one approach or the other is objectively mistaken, although that is possible.

Another way to look at the uncertainty in dose factors other than simply computing ratios of dose factors is to look at the variability around the ratios. To this end, we use the geometric standard deviation as a measure of variability, which can provide an estimate of the confidence that can be placed in any particular coefficient. The GSD of the ratios between draft NUREG-1640 and other studies amounts to a factor of 6 to 12,[11] which is a much larger range than the GSDs computed by draft NUREG-1640 based on its internal analysis of uncertainty (see Table 5-2). Although some difference would be expected, such a large discrepancy raises questions as to whether or not draft NUREG-1640's uncertainty bands are sufficiently wide to incorporate the range in which experts may reasonably disagree and therefore the bands might need rechecking. At the very least, the USNRC should understand and be able to explain the reasons for the discrepancy.

Given the complexity of the scenarios, the committee believes that an order of magnitude difference in dose estimates is reasonable for risk estimates of this type. With so much effort having gone into these studies over the past 20 years, it seems unlikely that additional, reasonable effort will be able to reduce dramatically the uncertainty in the coefficients that differ by less than a factor of 10—at least until there is real-world experience that can be used for benchmarking purposes. On the other hand, for the dose factors that show unusually large differences it would make sense to mount an international benchmarking exercise, with the goal of trying to understand the technical reasons for the major disagreements.

On average, the dose factors in draft NUREG-1640 and EPA TSD 97 will yield more restrictive secondary standards (i.e., the derived allowable activity level for release of a contaminated material will be lower) for the same primary dose standard than will the dose factors from the IAEA and EC studies. In other

---

[11] For instance, the committee looked at the GSD of the ratio of NUREG mean dose factors to those computed by the EPA and the EC (volume-contaminated metals), using data combined from Table 2.4 and Table 2.5 of NUREG-1640. The GSD was 6. A similar analysis was done for the ratio of NUREG-1640's mean dose factors to those computed by the IAEA (all materials), this time using Table 2.6 of NUREG-1640. The GSD was 8 for volume contamination and 12 for surface contamination. Note that 1 standard deviation is equal to the product of the median times the GSD; 2 standard deviations (~95th percentile for a log-normal distribution) equal the square of the GSD.

TABLE 5-3 Comparison of Dose Factor Estimates Made to Support Clearance Proposals

Category	USNRC NUREG-1640	EPA TSD 97
Nuclides	85	40
Scenarios	79	37
Approach	Generic geometries	Specific situations
Materials	Fe, Al, Cu metals; concrete, equipment	Fe, Al metals (copper in preparation)
Dose criteria	None established, estimates included for 10 μSv/yr when comparing results of other studies	None established
Exposed population	Member of critical group	Reasonable maximally exposed individual
Conversion coefficients	Traceable	Traceable
Collective dose considered	No	Yes
Comparison to fluxes from NORM or NARM, case-by-case clearance[c]	No	No
Dose uncertainty	Monte Carlo, traceable	Sensitivity studies and judgment
Level of conservatism in dose calculations[d]	Can be determined by policy maker	Implicit, thought to represent 90th percentile (e.g., 90% of members of critical group get lower doses)
Measurement uncertainty	Not considered	Considered in part
Human error	Not considered[e]	Not considered
Sensitivity studies	None	To determine which parameters contribute most to uncertainty
Benchmarking or validation	None	None

NOTE: NA = not applicable; NARM = naturally occurring and accelerator-produced radioactive material; NORM = naturally occurring radioactive material.

[a]IAEA (1988, p. 10).
[b]IAEA (1996, p. 47).
[c]To provide perspective.

EC-89, EC-113, EC-114	IAEA TECDOC-855	ANSI/HPS
104	56	52
Limiting pathway	NA	NA
Specific situations	Most conservative of dose factors from range of studies considered reasonable	Most conservative of dose factors from range of studies considered reasonable
Metals for recycle, buildings or rubble, all solids, equipment reuse	All solids	All solids
10 ∝Sv/yr; 1 person-Sv collective dose per year or, if higher, optimization (ALARA); skin dose of 50 μSv/yr	10 μSv/yr; 1 person-Sv per year or optimization (ALARA)[a]	10 μSv/yr; higher on a case-by-case basis.
Member of critical group	Member of critical group	Unspecified
Traceable	Traceable for volume-contamination factors	Not traceable
Yes	Yes[b]	Qualitative discussion
No	In part	No
Not formally analyzed	None	Assessed on an overall basis, not nuclide by nuclide
Implicit	Implicit	Implicit
Not considered	Not considered	Not considered
Not considered	Not considered	Not considered
None	None	None
None	None	None

[d]Dose calculations that result in higher percentile-valued dose factors are more conservative. NUREG-1640 reports a distribution of values and hence the selection is at the discretion of the policy maker.

[e]The USNRC has commissioned a separate study dealing with accidents.

TABLE 5-4 Ratio of NUREG-1640 Dose Factors to Other Estimates, Averaged Across Radionuclides

	"Mean" Ratio[a]	
	Volume Contamination	Surface Contamination
EPA metals[b]	0.64[c]	NA[d]
EC metals[e]	5.4[c]	10[f]
IAEA all materials[g]	14[h]	4.5[c]

[a]Computed as the exponential of the average of logarithms of ratios. The values from NUREG-1640 are all mean values that lie between the 50th and the 95th percentiles for all radionuclides.
[b]Derived from Table 2.4, draft NUREG-1640.
[c]±~26 percent. Standard deviation for an individual radionuclide, however, is approximately a factor of 5.
[d]Not applicable.
[e]Derived from Table 2.5, draft NUREG-1640.
[f]±37 percent. Standard deviation for an individual radionuclide is a factor of 12.
[g]Derived from Table 2.6, draft NUREG-1640.
[h]±36 percent. Standard deviation for an individual radionuclide is a factor of 8.

words, the draft NUREG-1640 and EPA dose factors are more protective. The committee has not been able to determine the precise reason for the differences from other estimates. The question of whether the total uncertainty could be greater on average than a factor of 10 is discussed in the next section.

Usefulness and Quality of Dose Factors

The committee's review of the studies listed in Table 5-1 found that some of the dose factors estimated in these studies, particularly those for radionuclides causing external gamma radiation doses to workers, can easily be shown to be reliable. Other dose factors require the use of parameters that are highly uncertain. One way to compensate for uncertainty in setting a protective standard is to set the dose factor for each radionuclide at a fixed margin above the best estimate for the dose factor. This allows the decision maker to compensate for the lack of complete knowledge in the dose analysis and thus increase confidence that the dose to the critical group will be below the primary dose standard. For example, the value for the dose factor can be set to the 95th percentile in the distribution of values for that dose factor rather than the median. Taking the *mean* value of the distribution, in almost all complex dose analyses (i.e., for right-skewed distributions), will increase the value of the dose factor over the *median* or 50th percentile result of the Monte Carlo calculation. The mean value has the property, in most calculations of this type, that its distance above the median automatically increases when uncertainty is large and decreases when uncertainty is small. (Although NUREG-1640 gives explicit values for the 5th, 50th, and 95th percentiles, it would be possible for the authors to extract other values—e.g., the 85th percentile—from the computed Monte Carlo distributions that would exceed the median by varying amounts.) However, the choice of any percentile level (and its corresponding dose factor), like the choice of a primary dose standard, is a matter of policy that cannot be decided by scientists through analysis or facts alone. For instance, policy makers could decide to choose dose factors closer to the median of the distribution of dose factors—forgoing the additional margin of protection afforded when a higher percentile-valued dose factor is selected—because they consider a 1 mrem/yr dose to be too far below background to be of concern. Conversely, they could pick a higher percentile-valued dose factor (e.g., the 95th) to assure the public that doses are very unlikely to exceed 1 mrem/yr.

If this additional margin of protection (which is implicit in the choice of higher percentiles) is not used in setting a dose factor, one must either pay close attention to the uncertainty in the estimate for each dose factor or fall back on assurances that analysts tended to be protective of public health (i.e., they picked parameter values—e.g., landfill leaching rates, resuspension coefficients—from the range of uncertain values that would end up being restrictive on the amounts of radioactivity that could be released to produce a given dose). However, the committee is reluctant to recommend reliance on statements by experts about the protectiveness of calculations. This is just the area in which experts have been found to perform poorly (Cooke, 1991; Shlyakhter and Valverde, 1995).

Although picking the percentile value appropriate for selection of dose factors is a policy choice, decision makers need to be informed about the quality of the supporting information. Over time, risk analysts have devised ingenious ways to reduce what at first glance appear to be unavoidable uncertainties in an analysis. For instance, it is often not necessary to know the *amount* of radioactivity released by a licensee in order to make use of a dose factor; often knowing the *mass concentration* is enough (i.e., the activity per gram). Analysts often simply consider releases that are large enough to saturate the doses to members of a candidate critical group, such as an entire truckload or industry-wide totals.[12] In general, bounding assumptions are made to eliminate the need to consider the total quantity of material released. Although this tends to overestimate dose factors and reduce allowed release concentrations, such as when concentrations are kinetically limited, it simplifies regulatory considerations. However, there are exceptions,[13] and some residual assumptions may still be necessary, such as the amount of mixing that takes place with nonradioactive material; see Box 5-2 for

---

[12]For example, once the volume of cleared material exceeds a truckload, the dose to the truck driver during one trip cannot go higher, which allows the number of trips one driver can make before receiving the allowed dose to be computed. The number of drivers needed to move the cleared material will increase as the quantity of cleared material increases, but this affects only the number of drivers who receive the dose. The collective dose increases, but not the dose to an individual driver.

[13]Exposure of workers in a steel plant may depend on the total quantity recycled (IAEA, 1992, p. 54), although even there, the dependence is limited. In the IAEA study, a hundredfold increase in the total amount of contaminated steel being handled produced an eightfold increase in individual dose (IAEA, 1992, p. 58).

**BOX 5-2
Computing Doses to Critical Groups
After Conditional Clearance for Landfill Disposal**

Suppose a secondary standard for conditional clearance of volume-contaminated materials for a single radionuclide (assume ^{137}Cs as an example) is set at 40 Bq/g and the conditions for release allow for landfill disposal. Draft NUREG-1640 assumes that mixing of the released material at the landfill with nonradioactive wastes is such that the released material constitutes only 0.15 percent of the volume in the landfill. Doses from gamma radiation exposure to persons living near the facility or playing golf on top of the landfill after it is closed can be computed from these starting assumptions. If one also assumes (or estimates from data) the rate at which the radioactive component is leached from the material into subsurface moisture, the partition coefficients and flow rates for transport of the radionuclide plume through the unsaturated zone to groundwater, and the direction and rate of groundwater flow, maximal doses can be estimated for persons drinking water from wells in the vicinity of the landfill or from surface waters fed by the groundwater.

These estimated doses to persons affected by the landfill can be compared with the computed dose to truck drivers who transport released material to the landfill. By comparing the doses to individuals from each exposure scenario (a materials truck driver, a golfer, a local resident drinking well water, a city resident drinking water from a downstream reservoir), an analyst can determine which category constitutes the critical group. This comparison among exposed groups to identify the critical group depends only on the concentration of the radionuclide in the material, the dilution factor, and possibly the size of the landfill, not on the total amount of radionuclide in the landfill. Nevertheless, despite this insensitivity to total amounts of radioactivity, considerable uncertainties may remain when it comes to estimating water contamination. Leach rates and the parameters for subsurface transport can vary enormously from default values.

an illustration. In contrast to individual doses, collective doses under a clearance standard are directly related to the total amount of radioactivity released. Despite the use of bounding assumptions, considerable uncertainty remains in some scenarios, particularly when it comes to predicting the behavior of radioactive materials leaching from landfills.

Analysts often add margins of protection to components of a dose factor calculation because information about a parameter is lacking or because the analyst is seeking greater generality for the analysis. Because different analysts may not use the same margins in their computation, the various studies listed in Table 5-1 are difficult to compare. The numbers are neither pure "best estimates" (i.e., estimates of central tendency) nor pure bounding estimates (estimates of the upper and lower bounds of a percentile range). It is particularly difficult to estimate how the dose factor calculated for one study would change if an assumed margin of protection were changed to improve its agreement with other studies.

Until a clearance system is implemented and concentrations of radioactivity in key scenarios are measured, one cannot be certain that assumptions made to provide margins of protection or other safety-enhancing factors have been adequate or are unrealistically restrictive. One way to deal with hypothetical model error is to adopt a policy of "adaptive management" in which real-world performance is monitored through validation that is possible only after implementation, or through retrospective analysis of selected case studies.

For example, leachate can be sampled from representative landfills, or concentrations of radioactivity in sample pieces of recycled steel can be checked, to ensure that the model assumed in calculating dose factors reasonably represents reality, with an adequate margin of protection. The model, and the dose factors calculated from it, should be updated if the primary dose standard is being exceeded or even if key assumptions in the model are clearly inadequate.[14] The IAEA encourages this type of retrospective review, including the "testing of radioactive consumer products on the market" (IAEA, 1988, p. 14). Reaching most of the limiting conditions that were assumed in estimating dose factors, such as truck drivers handling slightly contaminated truckloads every work day or concentrations in landfills reaching the maximum capacity, will sometimes take considerable time (typically, years of activity after clearance standards are implemented). If a validation program is in place soon after a standard is implemented, there will be sufficient time to adjust dose factors (and the clearance standards derived from them) if corrections are needed.

Based on Table 3-7, it seems unlikely that SRSM from USNRC-licensed facilities cleared under a dose-based standard will come close to matching the

---

[14] A validation program might also include measuring the distribution of radioactivity, or limits on the amount of radioactivity, that arrives at monitored landfills. Even data from portal monitors placed at both the sending and the receiving facility would be useful, particularly in assessing how often human error leads to gross errors in maintaining transport constraints. Other ideas for useful data collection can be gleaned from the EC guidelines, which require licensees to track the total amount of material cleared for disposition (EC, 2001). If the amount of material per shipment was recorded, as well as the activity measurements made to check compliance with the secondary standard, then uncertainty margins relative to the assumption of clearance at the activity level of the secondary standard could be computed. This analysis would also aid in determining if significant mixing of waste was occurring.

The USNRC may not find it justifiable to require this degree of data gathering and reporting by licensees, but it might fund a program of research-oriented activities. During the 1980s, when the Low-Level Radioactive Waste Policy Amendments Act was passed, the USNRC considered including a requirement in 10 CFR Part 61 for reporting data on the radioactive content of low-level radioactive waste shipments to disposal sites. This requirement was not included in the rule. It was believed that the data would be useful only as a broad check of assumptions made in the environmental impact analysis for disposal, not for material balance. For some years, such data were obtained by contract for such a broad check (personal communication from Robert Bernero, Board on Radioactive Waste Management, National Research Council, July 17, 2001). However a detailed material balance would not be necessary for the validation activities discussed.

concentration and total amounts of naturally occurring and radioactive material (NORM), naturally occurring and accelerator-produced radioactive material (NARM), and comparable materials that are cleared today under a case-by-case approach. Consequently, field data will probably prove useful only in assessing how well the clearance models have bounded the concentrations and thus estimated the doses. Nevertheless, a modest monitoring effort would boost confidence in the dose factors, particularly for those who are skeptical of the models being used. It may also provide useful incidental information on where NORM and NARM are ending up.

General Limitations of the Reviewed Studies

Failure to Consider Uncertainties Associated with Implementation of a Primary Dose Standard

Dose factors as estimated to date are useful theoretical tools. However, they have practical value only within a specific implementation protocol, where such a protocol can introduce uncertainties into dose estimates tied to primary dose standards. Only a few studies (e.g., EPA, 1997a) appear to have explicitly considered any implementation issues in assigning uncertainties to the estimated dose factors. Among these sources of added uncertainty are averaging error, sampling error, rounding error, and treatment of multiple radionuclides:

- *Averaging error.* The area or volume over which one averages radioactivity can introduce errors (EC, 2000, p. 20). This will increase the uncertainty associated with dose estimates.
- *Sampling error.* Guidance for a volume contamination standard would probably include acceptable sampling and modeling methods, which would allow some level of sampling error. Sampling error, in turn, could add to overall dose uncertainty. To a degree, any error incurred from a finite number of samples might be offset by the fact that not all of the cleared material will have an activity level exactly matched to the secondary standard. On the other hand, there is also the possibility that hot spots may have been missed.
- *Rounding error.* For practical reasons, regulatory authorities may decide to round secondary activity standards to a few convenient values—for instance, 0.1, 1, 10, and 100 Bq/g, and so forth. This can result in an error of a factor of three or so in dose factors. This practice, which has been adopted by the European Union (EC 2001, Table 1) and is used in the ANSI/HPS standard (ANSI/HPS, 1999), is equivalent in effect to choosing higher or lower percentile-valued dose factors. The possibility of rounding the derived secondary standards to integral powers of 10 should be considered when assessing uncertainties and selecting the percentile

value corresponding to the dose factors. The percentile level implicit in a rounded activity standard should be roughly the same as the percentile level sought in a dose factor that will not be subject to rounding. For example, if the policy choice for selecting dose factors is to maintain a 95 percentile level, then the implicit percentile level of a rounded activity standard should be at least 95 percent. Alternatively, information such as the implied confidence level after rounding should be presented with the proposed activity standards so that policy makers understand the implications of adopting a policy of rounding the activity standards.

- *Multiple radionuclides.* If a dose-based clearance standard was chosen, a decision would have to be made on whether its implementation for multiple radionuclides should apply a sum-of-the-fractions[15] computation or apply the individual clearance levels for any nuclides detected. The sum-of-the-fractions method is used routinely for control of radioactive effluents (10 CFR Part 20, Appendix B) and is recommended by the EC for clearing solid material (EC, 2001, p. 14). For a given protocol, an analyst can estimate the uncertainty that may result from using it with contamination from multiple radionuclides and include the estimated uncertainty in the dose factors. Without the specification of a protocol for treating multiple nuclides, it is difficult to assess whether any changes need be made, up or down, to the uncertainty estimates for dose factors.

Lack of Validation of Model Estimates

Validation against field data provides the best way to check for model error, as well as unexpected problems with parameter assignments. As noted in the previous section, a validation program should be used to correct and refine a system of dose-based clearance standards, given the inevitable uncertainties in the process of estimating dose factors. Furthermore, the confidence of policy makers, licensees, the public, and skeptics in the predictions from risk assessments can be increased by undertaking validation exercises. The committee heard only one presentation about a study in which clearance model estimates have been field-tested.[16] In that case, an international group led by the Swedish Radiation Protection Institute attempted to check predictions of model estimates against

---

[15]A sum-of-the-fractions computation is used when the governing standard sets the amount of each isotope that, if alone, would reach the dose limit of the standard. When materials containing many isotopes are analyzed for compliance with the dose-based standard, a fraction is calculated for each isotope present (the amount detected divided by the dose limit amount set for that isotope). The sum of all these fractions must be less than or equal to 1 if compliance with the dose limit is to be ensured.

[16]Shankar Menon, program co-ordinator, OECD/NEA Co-operative Program on Decommissioning, presentation to the committee, June 13, 2001.

results of actual recycling of SRSM. The committee did not review this work but wishes to encourage that such studies be undertaken.

Lack of Inclusion of Accidents and Human Errors in the Dose Factors

The IAEA recommends consideration of accidents in estimating exposures of the public from disposal exemptions (EC, 2000, p. 20). Examples of human error that can initiate or contribute to accidents involving error in clearance of materials at a nuclear power plant include failure to monitor properly, failure to properly handle and contain loose contamination, and delivery of material to the wrong recipient. Specifically, a facility that was routinely required to screen all scrap material for radioactivity, but rarely encountered any contamination, might disable the radiation alarms, fail to keep them in working order, and/or ignore them when they actually went off. Human error was not explicitly addressed in the analyses supporting dose factor estimates in any of the studies reviewed. However, the USNRC has carried out an (as yet unpublished) analysis of one form of human error (accidents), which suggests that this type of human error is not likely to have a significant impact on dose factor estimates. USNRC staff were not able to provide the study committee with the frequency at which exit monitors at licensed facilities were triggered by shipments on their way to final disposition, following clearance based on Regulatory Guide 1.86, a license provision, or approved by case-by-case review. However, a health physicist from the steel recycling industry told the committee that shipments from USNRC-licensed facilities have been sent back from recycling facilities because the shipments triggered portal monitors. Although alarm events could be false alarms since the portal monitors are set as close as possible to background radioactivity levels, they may also indicate that human errors were made in the release of material from the source facility. Consequently, it must be presumed at this time that some shipments will leave licensed facilities with contamination in excess of a clearance threshold level. Clearance coefficients that are estimated using a probabilistic approach, such as draft NUREG-1640, can account for this possibility.[17]

Human error may have only limited impacts on dose factor estimates, especially for those coefficients where simplifying methods have been used to make

---

[17]If human error is not correctly accounted for in the dose rate coefficients themselves, other methods can be used to handle it in the system itself. For instance, portal monitors can be placed not just at the exit of licensed facilities, but at recipient sites, such as landfills or recycling facilities, if release to these facilities is allowed under the standard adopted. In many cases, steel mills have such portal monitors (and, in some cases, monitors in other portions of the facility), as do landfills and licensees that generate wastes. Pennsylvania already has a requirement that all landfills be outfitted with portal monitors to catch orphan sources, along with a formal plan for dealing with radiation sources that trigger the monitors. As one example: if landfill disposition of SRSM were restricted to landfills that installed portal monitors, one protection against human errors made at licensed facilities might be institutionalized.

estimation easier and more robust. However, human error can also be embedded in the larger framework of system failure, which includes the following interrelated sources of failure: (1) hardware, (2) software, (3) organizational, and (4) human (Haimes, 1991). A follow-up study might take such a systems approach.

Potential Inconsistencies in Dose Factors Between Countries

As noted above, analysts from different countries have estimated different dose factors. These differences can lead to inconsistencies between clearance policies adopted in different countries. However, in discussing transnational consistency of dose factors and derived secondary clearance standards, two types of consistency must be distinguished. If countries agree on the same primary dose standard, they have agreed on the level of risk that sets the ceiling on clearable SRSM. For instance, there is widespread agreement on a 10 mSv/yr primary dose standard in the European Union. If countries disagree on which sets of dose factors are appropriate—the second type of inconsistency—they are differing over technical calculations, possibly differing only over degrees of conservatism that are needed to simplify the estimates.

Consistency of clearance standards across national boundaries is clearly desirable, particularly for materials that might find their way into international commerce. However, it would be inappropriate for one country to change its view of the supporting scientific evidence simply to achieve consistency with the standards in effect in other countries. Such an approach would not be conducive to building confidence in the scientific and engineering foundations for clearance standards.

Even the appearance of making changes in technical documents to make policy choices easier could undermine public confidence in the overall results. If rationalization of standards across borders becomes paramount, after attempts at technical rationalization have failed, the effort should be separated from the scientific deliberations by which dose factors are estimated. The decisions to rationalize for reasons beyond those supported by technical studies should be made as a clear policy choice (e.g., accepting more or less conservatism in the adopted dose factors).

DETAILED COMMENTS ON NUREG-1640

As noted, the committee paid particular attention to the draft NUREG-1640 because it was prepared for the USNRC in preparation for reconsideration of clearance standards. The discussions in this section supplement the earlier general discussion of analytical limitations in the draft document. Many of the issues raised here may have been considered intuitively by the analysts and staff that prepared the draft and judged to be of little consequence. Some may be currently under study at the USNRC. In any case, the committee believes that all of the following issues have to be considered explicitly, at some point, in the technical support process.

Issue 1: Landfill Disposal Scenarios

Landfill issues in the draft NUREG-1640 were difficult to understand. They require clarification and justification. The following are examples:

- *Fraction of material that goes to landfill.* The justification for the assumed 0.15 percent fraction of volume of material that ends up in a landfill is weak (USNRC, 1998b, p. 4-98). The ±50 percent uncertainty assigned to the fraction seems small.
- *Alternative economic models for landfill deposits.* Draft NUREG-1640 does not consider the situation in which only a small number of facilities are willing to take cleared material. Neither does the EPA, although TSD 97 does mention this possibility. If the postclearance landfill industry splits this way, the net result would be to increase the fraction of released material in the few facilities that would take contaminated material, thereby increasing the dose to landfill workers and nearby residents. This possibility is sufficiently realistic that it deserves assessment. It can probably be handled in draft NUREG-1640 by changing the uncertainty distribution currently assigned to landfill clearance calculations.
- *Uncertainties.* Landfill scenarios in draft NUREG-1640 did not have defined critical groups, so they did not get the consideration they might have if conditional clearance had been under consideration. Leaching rates, liner failure, and long-range transport are possible issues that should be addressed more carefully as part of the technical support process.

Issue 2: Incineration Pathway

Once material is released into general commerce, it may one day enter the municipal waste stream. Since a certain percentage of trash is incinerated to reduce volume, one possible immediate or delayed-clearance pathway would be incineration; yet this pathway was not addressed. Even though this pathway is unlikely to be significant, it should be explicitly considered.

Issue 3: Sensitivity Analysis

The uncertainty analysis was reasonable, but since the study uses a Monte Carlo analysis, the committee wondered why a set of sensitivity analyses was not carried out. Sensitivity analyses can be misconstrued as uncertainty ranges, but the committee believes that they can be constructive. Sensitivity studies yield important information about the significance of an input parameter's value to the output value predicted by the model. In this case, such a study would allow a better assessment of the effect of the parameter's uncertainty on the calculated dose factors. (See also discussion of resuspension of contamination below.)

Issue 4: Validation

There is no benchmarking or validation provided in the appendix material to draft NUREG-1640. Benchmarking or validation exercises would be appropriate to demonstrate the validity of the modeling technique.

Issue 5: Sample Calculations

There was a dearth of sample calculations that could have provided clarity for readers as to the overall method.

Issue 6: Multiple Pathways

The draft report does not consider multiple pathways. The committee notes that when exemption from regulatory control is considered, the IAEA (1988) recommended as follows: "The dose to the individuals of the critical groups(s) and the dose to the whole population exposed by the practice should not be significantly affected by other *similar* (or identical) practices (e.g. several waste disposal sites in the same region)."

Issue 7: Resuspension of Contamination

The draft document has only limited consideration of resuspension of surface contamination into the air. Of all the factors that can play a role in exposure to toxic substances, resuspension is probably the most difficult to address (IAEA 1992, p. 66; USNRC, 1998b, p. 3-8). Even after loose material is removed during cleaning, some residual radioactive material can be available for resuspension over a longer time. Resuspension rates, which generally affect only inhalation exposures, can span many orders of magnitude, as the authors of draft NUREG-1640 acknowledge: "The resuspension factor, RF_{sc}, is the most poorly known parameter in the inhalation pathway analysis . . ." (USNRC, 1998b, p. 3-8).

The method of uncertainty analysis adopted by NUREG-1640, which the committee applauds, can nevertheless be disconcerting when applied to parameters with large uncertainty ranges. The 95th percentile can end up being many times greater than the highest value measured to date. There is a tendency for analysts to disbelieve such numbers and make some form of downward adjustment. This is a potential form of downward bias that bears watching, given the known problem of expert overconfidence (Cooke, 1991), which leads to underestimation of uncertainty ranges when subjective judgments are made.

For example, in estimating doses to workers in reused trucks, the draft NUREG-1640 analysts selected the bottom of the range of resuspension values available to represent the median of the distribution, with little justification. The choice of geometric standard deviation was also made with little justification.

With measured resuspension rates varying by many orders of magnitude, it is difficult to determine how to handle this problem. At a minimum, a sensitivity analysis should be performed to inform readers as to how the dose factor would vary with a change in the resuspension coefficient.

A sufficient technical basis may not yet exist for assigning a credible uncertainty factor to certain types of releases that are sensitive to resuspension. If so, such clearance categories could be excluded by regulation until a sufficient technical basis is developed.

Issue 8: Collective Dose

Draft NUREG-1640 has no consideration of collective dose. The EC and the IAEA have a two-part primary dose standard, 10 µSv/yr for an individual and 1 person-Sv/yr for the collective dose to the population. Specifically, IAEA recommends that regulatory authorities conduct a generic study in the early stages of regulatory development to determine whether the annual dose from exempt practices will exceed one man-Sv. If not, then further optimization of the regulatory option being proposed is not needed (IAEA, 1988, pp. 10-11). Unlike practices from which individual doses may vary over a wide range and be a significant fraction of background radiation, doses from activities that result in low individual doses result in doses and therefore risks that are individually and collectively very small both in absolute value and in comparison to natural background and man made exposures—levels at which the significance of collective dose has been controversial. (To exceed the collective dose requirement, more than 100,000 persons would have to receive the allowed individual dose.) The EPA has also examined collective dose (EPA, 1997a). However, technical analysis by the USNRC has focused only on the individual dose. Although the dose to an individual in a critical group, and thus the secondary activity standard, does not ultimately depend on the total radioactivity released, the collective dose does increase with total radioactivity. For example, if more than one truckload of material is shipped at a given concentration from a licensed facility using different drivers for each truck, the dose to an individual driver from a full load does not increase, only the number of exposed truck drivers increases. Even if the same driver makes multiple trips, the dose will be limited by the total number of trips that can be made in one year.

Consequently, it may be of interest in shaping policy to have some idea of collective dose, recognizing that such estimates may carry much greater uncertainty than will the dose to an individual in the critical group. Given a collective primary dose standard in the range of 1 person-Sv/yr, the individual dose estimate for material from USNRC-licensed facilities is likely to be more restrictive than the collective dose (Clarke, 2001).

Issue 9: Size of Critical Groups

The draft NUREG-1640 does not discuss the total number of people exposed in any critical group. Although most critical groups will include a relatively small number of persons, other critical groups may include greater numbers of people. When groups are large, it is easy to think of smaller subgroups that could get higher doses. For instance, iron workers, train conductors, and elevator operators could receive higher doses from slightly radioactive steel than would users of common consumer objects. Knowledge of the approximate size of critical groups assists in building confidence that a more important subgroup has not been overlooked.

Issue 10: Total Activity Buildup and Mass Balance

The draft NUREG-1640 contains limited information on total activity buildup and mass balance. The methodology chosen to estimate doses for draft NUREG-1640 largely eliminates the need to know the total inventory of curies released. The authors consider (justifiably) that the total amount of curies released and stored affects the estimation of cumulative doses more than the estimates of critical doses (i.e., the individual doses on which dose factor selection is based). Nevertheless, the committee is uncomfortable with the lack of activity balance estimates.

Given that a material flow model has already been developed for the analyses, it should be straightforward to account for approximately how much of the radioactivity released each year is removed from the commerce "pool" as it enters landfills, how much will build up in the steel content, and how much would end up stored in structures. Since 85 percent of the steel cleared from USNRC facilities is likely to end up in landfills, steel made from cleared scrap will constitute only a tiny fraction of the total recycle in the United States. With radioactive decay, the committee does not believe that the buildup is likely to be significant, but without supporting estimates there is no explicit basis for this position.[18]

---

[18] It would also be useful to compare the amount of radioactivity in material projected to enter commerce and landfills from various proposed clearance policies with the amounts entering now from both the USNRC's case-by-case clearance policy and NORM or NARM sources. To aid in estimating the quantities entering commerce and landfills now from USNRC-licensed facilities, analysts could collect a random sample of case-by-case decisions from each USNRC region and analyze the dose implications using NUREG-1640 coefficients.

Issue 11: Accounting for Human Error

Accounting for human error is good risk assessment practice. Draft NUREG-1640 does not consider human error and specifically assumes that there is none. Although USNRC staff has already taken steps to analyze the impacts of accidents on dose factor estimates, more of this type of analysis should have been done in the draft document.

For instance, in one case, the analysis assumes that loose surface contamination is always removed according to good health physics practice (USNRC, 1998b, p. 3-2).[19] Yet inclusion of a modest human error rate could end up dominating the dose estimate. It is inconceivable that all loose surface contamination will always be removed prior to clearance. The probability that loose material may be overlooked may be low, but the downstream dose from loose contamination could in principle be sufficiently high to overcome the low probability that an error will occur.

Issue 12: Uncertainty in Conversion between Intake and Dose

The authors of draft NUREG-1640 did not consider uncertainties in the coefficients that convert inhalation and ingestion to dose,[20] relying instead on coefficients developed by the EPA. Although the uncertainty in these coefficients may not be significant compared to other uncertainties that enter the estimate of dose factors, especially for USNRC-licensed facilities,[21] this contribution to uncertainty should be explicitly considered.

FINDINGS

Finding 5.1. Analytical work in the United States and abroad over the past two decades is useful in understanding the likely doses associated with exposure scenarios that might occur under various clearance standards. Much of the technical analysis in this field has the objective of understanding "dose factors," which to date have been analyzed in depth only for (unconditional) clearance scenarios. A dose factor is used to convert a concentration of radioactivity that is about to be released, whether it be confined to a surface or contained within a volume, to a primary dose level (measured in microsieverts per year or millirems per year). With such a dose factor in hand, a primary dose standard can be converted to obtain a secondary clearance standard in terms of radionuclide activity, which

---

[19]"Based on assumed good health physics practices at NRC licensed facilities, removable surface contamination has been removed during decontamination procedures prior to final survey and clearance" (USNRC, 1998b, p. 3-2).

[20]Constant values taken by draft NUREG-1640 included "the dose equivalent due to radionuclide intake."

[21]The radionuclides of significance at USNRC-licensed facilities are generally not transuranics.

could then be used at USNRC-licensed facilities. A dose factor can be used with any choice of primary dose standard.

Finding 5.2. Selecting a primary dose standard is a policy choice, albeit one informed by scientific estimates of the health risk associated with various doses. For instance, as shown in Table 1-2, a lifetime dose rate of 10 µSv/yr (1 mrem/yr) equates to an estimated increased lifetime cancer risk of 5×10^{-5}, which falls within the range of acceptable lifetime risks of 5×10^{-4} to 10^{-6} used in developing health-based radiation standards other than radon in the United States (NRC, 1995, p. 50). When setting primary dose standards, regulators can make a policy decision to include a level of conservatism such that the final standard is in excess of the best-estimate dose factor and in this way account for uncertainty (e.g., selecting the 90th, 95th, or other percentile in the distribution for the dose factor, instead of the best-estimate value).

Finding 5.3. The uncertainty in dose factor estimates is a key technical issue. When an uncertainty has been estimated, a quantitative determination can be made of the likelihood that the dose to an individual in the critical group will be below the primary dose standard. Quantitative uncertainty estimates can also assist regulators in assigning a level of conservatism to dose factors in excess of the best estimate. Dose factors developed by analysts from different countries show wide variation, which highlights the need for careful consideration of uncertainties.

Finding 5.4. The committee concludes from its review that of the various reports, draft NUREG-1640 (USNRC, 1998b) provides a *conceptual framework* that best represents the current state of the art in risk assessment, particularly with regard to its incorporation of formal uncertainty, as judged using recommendations of this committee and other committees of the National Research Council. Once the limitations in draft NUREG-1640 have been resolved (see Findings 5.5 and 5.6) and the results are used in conjunction with appropriate dose-risk estimates—in the final version of the report or in follow-up reports—the USNRC will have a sound basis for considering the risks associated with any proposed clearance standards and for assessing the uncertainty attached to these dose estimates.

Finding 5.5. The development of the NUREG-1640 draft has been clouded by questions of contractor conflict of interest. The mathematics and completeness of scenarios considered in draft NUREG-1640 have been verified through an audit carried out by another USNRC contractor. The committee also carried out its own review that generally confirmed the reasonableness of several dose factor analyses. However, a thorough review of the choice of parameters and parameter ranges, term by term, is needed to complete the reassessment of draft NUREG-1640.

Finding 5.6. Draft NUREG-1640 did not consider human error and its possible effect on dose factor predictions, nor did it consider scenarios involving multiple exposure pathways. In addition, draft NUREG-1640 does not provide a sufficient basis to analyze conditional clearance options, such as disposal in a Subtitle D landfill.

Finding 5.7. The dose factors developed in draft NUREG-1640 should not be used to derive clearance standards for categories of SRSM other than those considered in the draft NUREG-1640, without first assessing the appropriateness of the underlying scenarios. Some of the dose factors developed in draft NUREG-1640 are likely to require modification when applied to other mixtures of radionuclides (e.g., mixtures in which transuranics dominate) and other clearance scenarios, such as may be relevant to DOE material and technologically enhanced naturally occurring radioactive material (TENORM).

6

Measurement Issues

The quantitative determination of the identity and activity of radionuclides present in a sample is a process that ranges from straightforward to complex, depending on the radionuclides, their distribution on or within the sample, the instrumentation available, the material matrix, and the pattern of radionuclide distribution within the matrix. Many radionuclides that emit gamma photons are relatively easy to identify and quantify. Most radionuclides that decay only by particle emission can be detected if they are on the surface of a solid material, but identification of the specific radionuclides present is often difficult or complex. Further, when particle-emitting radionuclides are distributed through the volume of a solid material, determining the amount of a radionuclide(s) present can require sophisticated technology beyond simple survey instruments.

Dose cannot be measured directly. Instead, the dose received is estimated by first determining activities for the radionuclides to be released (identity and quantity of each radionuclide) then using a factor to convert from activity to dose. Specifically, a screening level of activity is set by two quantities, the primary dose standard and the dose factor that relates the secondary activity standard (or screening level) to the primary dose standard, as discussed in Box 5-1. The dose factors, which are derived by modeling, vary by radionuclide and by the expert group that computed them. The relationship between source concentration and dose is affected by many factors, including but not limited to the following:

- The magnitudes of the dose factors chosen to derive screening levels from the primary dose standards;
- The specific instrumentation used in measuring radioactive material concentrations in a source;

- The counting conditions, including background radiation levels;
- The sample's physical and chemical characteristics;
- The inventory (identity and quantity) of the radionuclides present; and
- The nonradioactive material present.

NUREG-1507, *Minimum Detectable Concentrations with Typical Radiation Survey Instruments for Various Contaminants and Field Conditions* (USNRC, 1997) discusses each of these factors in detail, including the factor's impact on the minimum detectable concentration (MDC). The MDC is defined in NUREG-1507 as "the minimum activity concentration on a surface or within a material volume, that an instrument is expected to detect (e.g., activity expected to be detected with 95% confidence)" (USNRC, 1997, p. 3-1).

This discussion assumes that (1) the concentration of any radionuclides in samples to be measured is low relative to licensed levels and (2) the dose received by individuals from contact with these materials after their release is a small fraction of the natural background doses. As the activity in the sample increases, detecting, identifying, and quantifying the radiation source or sources become easier.[1]

When clearance for materials is considered, the process starts with an assay of a sample having an unknown inventory of radionuclides. The instrument selected to perform the assay will depend on the type of radiation that may be present. The *Multi-Agency Radiation Survey and Site Investigation Manual* (MARSSIM) (EPA et al., 2000) specifies a methodology, which is discussed later in this chapter, for accomplishing a statistically valid assay of radioactivity in potentially clearable material. It also provides guidance on instrument selection. NUREG-1507 provides detailed information on instrument capabilities (USNRC, 1997).

Instrument selection is straightforward when it is known which radionuclides could be present. An example would be a medical licensee that uses only three radionuclides. However, if the licensee operates a reactor where a large number of radionuclides are present and neutron activation of materials is a possibility, instrument selection may be more complex. A series of measurements may be required, using different instruments, each of which can detect a different radiation type. Each measurement will yield a number of counts obtained in a counting period. The counts per unit time are converted to units of radioactivity, using the known properties of the detector and the geometry of the configuration for counting (see Appendix E).

An important issue is whether one or more radionuclides may be present. Each radionuclide has its own activity, which in most circumstances will differ from the activities of other radionuclides present in the sample. However, as the

---

[1]Appendix E of this report provides tutorial-level information on radiation, radioactivity, and radiation detection.

number of radionuclides present increases, it becomes increasingly likely that the radiation from one will mask (be sufficiently close in energy to) the radiation from another, complicating the process of identifying and quantifying them.

Detection limits for both field survey instruments and laboratory instruments play a critical role in selecting the instrumentation and measurement procedures used in the analysis. Background radiation from naturally occurring radionuclides and cosmic radiation influence the sensitivity of the measurement process. As discussed in Appendix E and NUREG-1507, a detection limit in effect represents a practical trade-off between the acceptable statistical chances of obtaining a false positive or a false negative indication of the presence of radioactive material.

LEVELS OF DETECTABILITY

A reasonable question to ask is whether a radionuclide can be measured at the concentrations corresponding to (i.e., derived from) proposed primary standards.

The Environmental Protection Agency's (EPA's) Technical Support Document 97 ("TSD 97") presents MDC data derived from 24 laboratories (EPA, 1997a). The authors of TSD 97 recognized that increasing the count time or sample size could lower the reported MDC, but they concluded that the values reported represented the state of the art at the time (1995) for practical measurements. For most radionuclides, the background count rates were less than one count per minute and the lower limits of the detectors were less than 0.037 Bq/g (1 pCi/g). A variety of instruments were used, depending on the radionuclide. Count times ranged from 20 to 1,000 minutes. Sample masses ranged from 0.1 to 750 grams.

A review of the dose factor data illustrates the wide range of screening levels for volume contamination (picocuries per gram) obtained from different reports for the same radionuclide. Table 6-1 presents the screening levels for selected radionuclides from three reports, based on a 1 mrem/yr primary dose standard. In the two right-hand columns are the volumetric MDCs from TSD 97. Despite the variations, these derived (secondary) screening levels[2] are all greater than the lower MDC from TSD 97, except for the ^{129}I dose factor for NUREG-1640. Even this screening level could probably be detected if longer counting times were used to lower the MDC. Thus, under practical measurement conditions, existing measurement capabilities are sufficiently sensitive to meet almost all of the de-

---

[2]Derived (secondary) screening levels (i.e., secondary dose standards) can be derived by dividing the primary standard (in units of microsieverts per year) by the highest dose, from the most critical scenario, per year per becquerel per gram for volume sources, or by the highest dose per year per becquerel per square centimeter for surface-contaminated sources (see Box 5-1).

TABLE 6-1 Comparison of Derived Screening Levels and Laboratory Minimum Detectable Concentrations (MDCs) for Selected Radionuclides (pCi/g)[a]

Radionuclide	Derived Screening Level				MDC	
	ANSI/HPS 13.12-1999	USNRC Values NUREG-1640 Table 2.6	IAEA TECDOC 855 Table 1.6[b]		EPA TSD 97 Table 8-9[b]	
			Low	High	Low	High
^{137}Cs	30	1	5.4	2,432	0.007	0.3
^{60}Co	30	1	13.5	2,432	0.01	0.3
^{63}Ni	3,000	27,000	2.15×10^5	2.7×10^7	1	100
^{129}I	300	0.1	270	21,000	0.4	2
^{14}C	3,000	17	2,700	1.9×10^5	0.2	37
^{239}Pu	3	1.2	2.16	18,000	0.02	0.4
^{99}Tc	3,000	2.3	1,100	1.6×10^6	0.3	15
^{230}Th	3	1.2	2.7	216	0.05	0.5

NOTE: ANSI/HPS = American National Standards Institute and Health Physics Society; IAEA = International Atomic Energy Agency; USNRC = U.S. Nuclear Regulatory Commission.

[a] Based on 1 mrem/yr.
[b] Low and high indicate the extremes of the screening level range presented in the reference.

rived (secondary) screening levels for volume contamination derived in the technical analyses reviewed by the committee.

TSD 97 also evaluated the detectability of surface contamination and reached a similar conclusion. Namely, existing measurement capabilities for surface contamination are sufficiently sensitive to reach the screening levels for surface contamination derived in these same technical analyses. Although the Health Physics Society (HPS) Standards Working Group evaluated a different set of instruments and measurement procedures for the American National Standards Institute (ANSI)-Health Physics Society Standard N13.12-1999, the conclusion about detectability at the derived activity levels was the same (ANSI/HPS, 1999, Sections B.4 and B.5):

> ... in most cases the minimum detectable activities were significantly lower than the derived screening levels. These results indicate that, with a careful selection of alpha and gamma spectroscopy instruments and methods, it should be possible to attain a minimum detectable activity lower than the screening levels for most groups of radionuclides identified in this standard.

The ANSI/HPS report uses the term "minimum detectable activity" instead of MDC.

MEASUREMENT ISSUES	119

TABLE 6-2 Detectability of Selected Radionuclides by Laboratory Analysis Relative to Derived Screening Level (DSL) from TSD 97 (pCi/g)[a]

Radionuclide	MDC Low	MDC High	DSL 15 mrem/yr	DSL 1 mrem/yr	DSL 0.1 mrem/yr	Detectable at All Levels?
^{137}Cs	0.007	0.3	170	11	1.1	Yes
^{60}Co	0.01	0.3	17	1.1	0.11	Yes
^{63}Ni	1	100	1.4×10^6	93,000	9,300	Yes
^{129}I	0.4	2	19	1.3	0.13	No
^{14}C	0.2	37	17,000	1,200	12	Yes
^{239}Pu	0.02	0.4	21	1.4	0.14	Yes
^{99}Tc	0.3	15	700,000	46,000	4,600	Yes
^{230}Th	0.05	0.5	23	1.6	0.16	Yes

[a]Low and high represent the extremes of the derived screen levels in this reference.
SOURCE: EPA (1997a, Table 8.9).

Table 6-2 compares the MDCs from TSD 97 with the derived screening levels from TSD97 for volumetric contamination corresponding to primary dose standards of 15 mrem/yr, 1 mrem/yr, and 0.1 mrem/yr. The scenario used to derive the screening levels was the normalized dose to individuals exposed to radiation as the result of recycling scrap metal from nuclear facilities. Again, the MDCs are lower than the screening levels in all cases except for ^{129}I at the 0.1 mrem/yr primary dose limit.

TSD 97 reports similar results for surface-contaminated materials, when large-area detectors are used for surface scans (EPA, 1997a, Table 8-6). For large-area detectors used in the scan mode with a distributed source, the TSD 97 analysis concludes that in the laboratory, all 40 radionuclides considered would be detectable at the surface contamination screening levels (in units of disintegrations per minute per 100 cm^2) derived from a primary dose limit of 1 mrem/yr. These results assume a scanning rate of one-third of the detector width per second for beta and alpha detection and 15 cm/s for gamma detection.

For small-area detectors, which TSD 97 assumes would be used in field conditions, detectability becomes more difficult when factors such as human error, small nonhomogeneous contamination areas, realistic distances from source to detector, the condition of the material's surface, and surface coating are included. The fraction of radionuclides detectable under field conditions at the derived screening levels decreases from 39 of 40 for a primary dose limit of 15 mrem/yr to 31 of 40 for 1 mrem/yr and only 11 of 40 for 0.1 mrem/yr.

Whenever the potential exists for the presence of radionuclides that are not detectable with the detection method being used for the survey, it is necessary to change or modify the method to increase the sensitivity of the measurement by lowering the scan rate, changing to a larger area detector, or changing from a field measurement to a laboratory measurement. The conclusion of TSD 97 is that at levels corresponding to the screening levels utilized in that study of 15 mrem/yr and 1 mrem/yr, "100% of the radionuclides evaluated can be detected." Even at screening levels corresponding to 0.1 mrem/yr, "85% of the radionuclides are detectable" (EPA, 1997a, p. ES-17).

Thus, for both volume-contaminated and surface-contaminated solid materials, measurement of radionuclide activity concentrations at levels being considered for dose-based standards is not the limiting factor if the primary dose standard is at or above 1 mrem/yr in both laboratory and field measurements.

MEASUREMENT COST

The cost of measuring activities at these levels depends on the difficulty of analysis. The instrumentation to perform alpha, beta, and gamma spectroscopy is similar in cost to the most sophisticated systems for chemical analysis. Alpha and beta spectrometers cost approximately $50,000 each, but many systems can be adapted to analyze either particle by changing the detector. Gamma spectroscopy systems range from $50,000 to $200,000. A reasonable cost to set up a state-of-the-practice radionuclide analysis laboratory would be less than half a million dollars. The major operating expense is for the trained personnel needed to perform the sample preparation analyses correctly, especially on difficult samples. The TSD 97 authors referenced an article by Cox and Guenther (1995) that presented a range of MDCs as reported by 24 commercial and governmental laboratories. Table 8-5 of TSD 97 presents detection costs in 1995 dollars per sample that range from $40 to $375, depending on the radionuclide. There is some increase in per-sample cost as the required sensitivity increases: activities in the 10 pCi/g range, cost $40 to $250 per sample; in the 1 pCi/g range, $75 to $300 per sample; and in the 0.1 pCi/g range, $100 to $375 per sample. However, the increase is not as large as would be expected if most laboratories offering detection services were not already working with instruments and measurement procedures adequate to detect activities at the 0.1 pCi/g level.

If clearance is an option, the tradeoff between the cost of clearance and the cost of disposal as low-level radioactive waste (LLRW) will ultimately determine which option a licensee chooses. Chapter 4 estimates that costs for LLRW disposal will range from $3,120 to $16,800 per cubic meter. LLRW densities in the United States are usually between 50 and 120 pounds per cubic foot (0.8 to 1.92 metric tons/m^3). If a nominal density of 75 pounds per cubic foot, disposal costs of $30 per metric ton and $110 per metric ton at Subtitle D and C landfills, respectively, and a fixed sampling cost of $20 per sample (collection and preparation) are assumed, one can estimate the number of samples that can be taken at the break-even cost relative to LLRW disposal. Table 6-3 presents the results of

TABLE 6-3 Estimated Number of Analyzed Samples per Metric Ton of Waste at Breakeven Between Clearance and Low-Level Radioactive Waste Disposal

Alternative Disposal Site	LLRW Disposal ($2,590 per metric ton) Number of Samples Analyzed at		LLRW Disposal ($13,950 per metric ton) Number of Samples Analyzed at	
	$40 per sample	$375 per sample	$40 per sample	$375 per sample
Subtitle D at $30 per metric ton	42	6	232	35
Subtitle C at $110 per metric ton	41	6	232	35

NOTE: Calculation assumes no difference in transportation costs and constant sampling costs of $20 per sample.

this estimation.[3] The number of samples required to characterize the waste stream adequately will depend on the degree of certainty that the waste is homogeneous. However, at the higher LLRW cost, the number of samples that could be taken for the same cost ranges from 35 to 232, which is large enough to characterize a homogenous ton of waste. If the lower cost of LLRW disposal and the high sample analysis cost are used in the estimation, the six samples at equivalent cost are probably too small for adequate sampling, unless the waste is known to be homogeneous. Depending on the waste stream and the sampling protocol, it may be possible to aggregate samples and resample. This approach would reduce the typical number of samples to be analyzed per ton of waste.

Thus, the cost of sampling and analysis by itself does not appear to be a limiting factor when selecting a primary dose standard at or above 0.1 mrem/yr. (However, as noted above, at screening levels corresponding to a primary dose standard of 0.1 mrem/yr, the detection capability of field instruments is such that only 11 of 40 key radioncuclides can be detected.) This conclusion on costs is confirmed by the operation of a commercial waste management service, Duratek, Inc., which uses the derived screening levels from ANSI/HPS Standard N13.12 (see Table 6-1) to make decisions on waste disposition. Duratek, Inc. provided

---

[3]For example, if you had one ton of waste and access to LLRW disposal at $2,590 per ton, one option is to send that ton of waste to such an LLRW disposal facility. On the other hand, to send the same ton of waste to a Subtitle D facility, you would have to sample sufficiently to show it meets clearance levels and do so within a budget of $2,590-$30 = $2,560, the amount left after $30 tipping fee per ton. At $40 per sample characterization plus $20 per sample for sampling, this allows 42 samples to be taken within the break-even budget. If more samples are needed to show the waste meets clearance criteria, it is cheaper to send it to LLRW disposal.

the committee with information on its process and procedures, as discussed in the next section.

CURRENT MEASUREMENT PRACTICES OF A WASTE BROKER

Radioactive waste is generated daily from hospitals, research laboratories, and nuclear power plants. Licensees that generate controlled materials during operations currently survey all potentially contaminated waste materials prior to shipment. Those that are determined to have no licensee-generated radioactive material present are treated as nonradioactive waste. Materials that have surface contamination are either treated as LLRW or cleared using the criteria in Regulatory Guide 1.86 (AEC, 1974), license conditions, or approval obtained on a case-by-case basis from either the USNRC or the agreement state regulator. During decommissioning, potentially radioactive materials are typically cleared on a case-by-case basis or sent to a waste processor for clearance. Known radioactive materials are disposed as appropriate for their radioactive waste classification.

In 2000, about 30,000 tons of LLRW were processed in the United States. Waste brokers and processors handle a significant fraction of this waste. Waste brokers provide services to direct the disposition of LLRW and to prevent the release of contaminated materials into general commerce. A broker may transport, collect, or consolidate shipments or process radioactive waste. The survey of the incoming waste stream is an essential step in a waste processor's management of customer materials. The incoming shipment is scanned with handheld counters as an initial screen. (The licensee shipping the material has already certified that the waste has a low activity level and can be evaluated for clearance.) The material is then examined in either a box or a drum assay system. At the facilities of the waste broker Duratek, Inc., high-purity germanium (HPGe) detectors are employed for gamma spectroscopy, sodium iodide detectors are used for micro-dose rate determinations, and the records for each assay are stored digitally. If the material is clean (no activity at or above detectable limits), it is shipped to a Subtitle D landfill. As a further check, portal monitors at the facility exits are used to ensure that "clean" material shipped to the local Subtitle D landfill will not trigger portal monitors upon arrival there. If the material is contaminated at levels above those that would allow landfill disposal, it is either returned to the generator or, at the direction of the generator, disposed of as LLRW. Prior to disposal as LLRW, material is processed by melting, compaction, incineration, or a combination of these processes, to reduce its volume (which reduces disposal costs).

THE MARSSIM METHODOLOGY

Determination of an appropriate sampling program is a major consideration in the measurement process. MARSSIM methodology could be a valuable tool for licensees in demonstrating compliance with the type of dose-based standards under consideration for releasing SRSM. The MARSSIM includes a statistical sampling methodology suitable for release of land and buildings potentially containing residual radioactive material in surface soil or on building surfaces. At some licensed facilities, potentially clearable building materials may contain volume-distributed sources of radioactivity, in addition to surface sources. The MARSSIM methodology could also be expanded to be used as a decision tool in evaluating these solid materials.

The number of measurements or samples needed in each survey unit for statistical testing of residual radioactive material against a release level depends on the expected variability in concentration of the radioactive material and the level of acceptable error. If a licensee is in doubt, MARSSIM encourages assuming a larger, rather than smaller, variability in the material. This conservative approach (presumption of less homogeneity) drives a MARSSIM-guided assessment toward taking a larger number of measurements or samples.

A plethora of radiation detection instruments is available to scan surfaces and make direct measurements of residual radioactivity. The radionuclide(s) present and the magnitude of the release level are key factors in determining the appropriate instrument for a particular slightly radioactive solid material to be assessed. Several references, including MARSSIM (EPA et al., 2000), and NUREG 1507 (USNRC, 1997), provide MDCs for various types of radiation detection instruments under different sets of circumstances. The characteristics of the detector (probe area, detection efficiency, background response, etc.) enable the licensee to relate the release level to a corresponding instrument response, which MARSSIM calls the Derived Concentration Guideline Level (DCGL). The instrument selected should have sensitivity as far below the DCGL as possible. MARSSIM recommends that the MDC should be less than 10 percent of the DCGL, although it is acceptable for the MDC to be as much as 50 percent of the DCGL.

Having selected appropriate instrumentation, the licensee must next develop an integrated survey design combining some degree of scanning surveys with static measurements or sample collection. MARSSIM strongly recommends that the effort expended be weighted toward those survey units[4] more likely to contain elevated levels of residual radioactive material.

---

[4] A geographical area consisting of structures or land areas of specified size and shape at a remediated site for which a separate decision will be made whether the unit attains the site-specific reference-based cleanup standard for the designated pollution parameter. Survey units are generally formed by grouping contiguous site areas with a similar use history and the same classification of contamination potential. Survey units are established to facilitate the survey process and the statistical analysis of survey data (EPA et al., 2000).

The assessment phase, which follows collection of the survey data, includes data validation as well as a reassessment of the quantity of data. For example, the number of measurements taken was based, in part, on an assumption about the variability of radionuclide concentration in the material. This assumption should be verified. If the variability was underestimated, more data should be collected to ensure that the desired statistical significance is attained. For survey units that are likely to contain elevated levels of radioactivity, MARSSIM also requires that an elevated measurement comparison (EMC) test be performed to demonstrate compliance for small areas with elevated activity concentrations.

FINDINGS

Finding 6.1. The concentration of radioactive material in released solids directly affects radiation detection requirements and costs. Measurement of the amount of radioactive material in a solid matrix is a complex task that involves a combination of instrument characteristics, background radiation levels, and source characteristics. No single measurement method would be appropriate or adequate for all radionuclides.

Finding 6.2. The overall measurement costs, including sampling (collection and preparation) and analysis and material disposition choices, affect clearance decisions. If the measurement costs are too high, it may be more cost-effective to dispose of the material as low-level radioactive waste.

Finding 6.3. For a 1 mrem/yr or higher standard (and the corresponding derived secondary screening levels), the majority of radionuclides can be detected at reasonable costs in a laboratory setting, under most practical conditions. For a 0.1 mrem/yr standard, the measurement capability falls below the upper bound of minimum detectable concentrations for some radionuclides in some laboratories, although 85 percent of radionuclides are still detectable. Using field measurements, a more rapid fall-off of detectability is observed at more stringent radiation protection levels, with 31 of 40 key radionuclides detectable at 1 mrem/yr and 11 of 40 detectable at 0.1 mrem/yr.

7

International Approaches to Clearance

THE GLOBAL CONTEXT

Import-export activities involving recycled materials have increased greatly with the growth of international trade over the past several decades. This is particularly true for metals such as steel in which recycled material constitutes a significant fraction of the total production. It is also true for metals with high intrinsic value such as aluminum, copper, and nickel. Scrap metal is actively traded worldwide, and the amounts in international trade are measured in millions of metric tons per year. The United States imports about 3 million metric tons of scrap steel per year. Both the European Union (EU) and the United States are concerned about imports of steel scrap containing radioactive material (see Box 7-1). The amount of scrap steel employed in making steel varies markedly with the process, but on the average, scrap represents a significant component of the charge for a furnace. The percentage of recycled material is also significant for some other metals such as aluminum, copper, and nickel. These high percentages reflect both the inherent potential for metals to be recycled repeatedly at a cost competitive with producing metal from raw materials, which is higher than for most other materials, and the actual practice in metals production worldwide.

Appendix D summarizes the work on slightly radioactive solid material (SRSM) clearance standards by various entities within the United States, as well as major international efforts. Specifically, Appendix D discusses the following documents developed by international organizations: (1) IAEA Safety Series 89; (2) EC Radiation Protection 89; (3) International Commission on Radiological Protection Publication 60; (4) reports of the United Nations Scientific Committee

> **BOX 7-1**
> **Sealed Radioactive Sources in Scrap Metal**
>
> One of the steelmakers' concerns is contamination of recycled materials due to the inclusion, whether accidental or deliberate, of sealed high-radioactivity sources in metal scrap for recycling. This possibility is a different issue from the introduction of slightly radioactive solid material cleared from licensed facilities. Cleared SRSM has presumably been properly evaluated and released according to approved criteria. Sealed sources that are either intentionally or inadvertently introduced into scrap offered for processing present a greater and typically unknown source of contamination. The management of such "orphaned sources" is beyond the scope of this report. It is mentioned here because the introduction of these sources typically dominates the discussion of recycle of radioactive materials into steel. These "orphaned sources" in metal scrap can contaminate a processing plant. Such contamination may raise questions regarding worker health in subsequent handling or processing of the scrap, as well as exposures to members of the general public during transport and any subsequent use of the contaminated metal.

on the Effects of Atomic Radiation; and (5) European Union Basic Safety Standards.

National and international concerns about potential problems of radioactive contamination associated with recycled metal have increased during the past decade. Several international agencies are addressing the problem, including the International Atomic Energy Agency (IAEA), the United Nations Economic Commission for Europe, and the European Commission (EC). At present, no international or national registries of missing radioactive sources are available to the recycling industry to indicate when such sources are lost or stolen and where they may enter the recycling chain. To address concerns about the import-export of metal scrap with undetected levels of radioactivity above clearance limits, the Team of Specialists on Radioactive Contaminated Scrap Metal, United Nations Economic Commission for Europe, has proposed the following (UNECE, 2001):

- The regulatory framework associated with the clearance of material should include provisions for prior notification to the receivers of the material of the origin of this material and the regulatory framework under which it is released.
- When materials contaminated with naturally occurring radioactive materials (NORM) are released according to a national regulatory framework, such information should also be forwarded.
- As part of the "contractual" provisions, this information should be conveyed with the released material to the successive suppliers and buyers of the metal scrap.

The European Union has been establishing standards and methods of control for SRSM within Europe. Many EU countries possess nuclear power reactors and nuclear fuel cycle facilities. As these facilities are decommissioned, scrap metals and concrete are cleared from regulatory control. Some of these materials are released for restricted uses, but others are released to general commerce. The amount of potentially clearable metal from all categories of EU facilities is estimated at 12,700 metric tons per year, although this estimate increases to about 40,000 metric tons by 2020 from commercial power plants alone (EC, 1998b).

Different clearance procedures for the release of SRSM metals are currently in use among EU countries. Delayed release and dilution have been standard practice in some. For example, 14,000 metric tons of contaminated steel scrap has been melted at a dedicated melting facility operated by Siempelkamp (Krefeld, Germany). Although most of this recycled scrap metal has been used in restricted applications, 2,000 metric tons has been released for unrestricted use. The contamination limits in Germany for unrestricted reuse are expressed in becquerels per gram for each radionuclide (e.g., cobalt-60 is 0.1 Bq/g).

The EU member nations are in various stages of developing detailed regulations to implement the controlling directive from the EU Council (EU,1996), as discussed in the next section. Japan is developing similar regulations and has ongoing discussions among government organizations. Table 7-1 summarizes international activities and the status of clearance standards for SRSM in a number of countries for which the committee was able to obtain information. Activities of the Department of Energy (DOE) and the U.S. Nuclear Regulatory Commission (USNRC) are included in Table 7-1 for comparison.

Generation of radioactive material outside the United States is not limited to EU member states or to commercial nuclear power operations and decommissioning. Nuclear weapons development has occurred in many countries over the past 60 years. China, India, and Pakistan are known to have developed and tested weapons. Clearly, radioactive materials containing significant quantities of long-lived radionuclides are located around the world.

Documentation regarding radioactive material contamination exists for republics of the former Soviet Union, which produced 55,000 nuclear warheads during the Cold War. The Soviet Union, and later Russia, produced uranium and plutonium for nuclear weapons at three closed atomic cities—Ozersk, Seversk, and Zheleznogorsk—which were founded to produce weapons-grade material and reprocess civilian nuclear fuel. Some of these materials may enter commerce as SRSM if cleared from one or all of these countries involved in the development of nuclear weapons, nuclear power, and other uses of radioactive materials in industry, medicine, and research.

For general information on radioactive waste management activities, the International Nuclear Societies Council (INSC) recently published an overview of radioactive waste management activities in countries with INSC member soci-

TABLE 7-1 International Clearance Status as of May 2001

Country	Surface Clearance Level(s) (Bq/cm^2)	Volume Clearance Level(s) (Bq/g)
Belgium	Case-by-case	Case-by-case
France	Nuclear power industry: moratorium on generic levels; case-by-case allowed Nonnuclear power industry: case-by-case	Nuclear power industry: moratorium on generic levels; case-by-case allowed Nonnuclear power industry: case-by-case
Germany	Nuclide specific, based on 10 μSv to a person in a year.	Nuclide specific, based on 10 μSv to a person in a year (e.g., 0.1 Bq/g ^{60}Co)
Japan	No general criteria	No general criteria

Basis for Clearance	Situation	Remarks
IAEA TECDOC-855[a] levels used as reference levels	General regulations are under review for update to Directive 96/29/Euratom[b]	IAEA TECDOC-855 dose criteria are 10 µSv to a person in a year, plus collective dose of 1 person-Sv or optimization
Waste stream analysis, quality assurance, impact study, presentation to public, specific authorization	Incorporation of Directive 96/29/ Euratom[b] for both power and non-power industries is in preparation, planned for mid-2001	Ministerial order issued Dec. 31, 1991, requested nuclear industry to implement waste stream analysis Authorized release is possible, though rarely used Generic clearance levels may be required for non-nuclear power very low level waste
SSK (Commission on Radiological Protection) recommendations.	Incorporation of Directive 96/29/Euratom[b] is in preparation Some debate on whether to replace SSK recommended levels with EC RP 122[c] clearance levels	Updated regulations targeted for fall 2001 Authorized release is possible (e.g., 4 Bq/g ^{60}Co for landfill or incineration; 0.6 Bq/g ^{60}Co for metals to be melted) Clearance of sites based on 10 µSv/yr. individual dose
Ongoing discussions among government organizations	Legislation targeted for 2001	Nuclear Safety Commission based clearance calculations on 10 µSv criterion. These agree well with IAEA TECDOC-855[a] with a few exceptions

continues

TABLE 7-1 continued

Country	Surface Clearance Level(s) (Bq/cm^2)	Volume Clearance Level(s) (Bq/g)
United Kingdom	Case-by-case basis	0.4 Bq/g for non-naturally occurring radionuclides Naturally occurring radionuclides range from 0.37 to 11.1 Bq/g, depending on the nuclide
United States	DOE suspension of scrap metal for recycling	DOE moratorium on metals
	USNRC: consistent with average of 0.017 Bq/cm^2 for transuranics, ^{226}Ra, and others to 0.83 Bq/cm^2 for most β-γ emitters	USNRC: no general criteria

[a]IAEA (1996).
[b]EU (1996).
[c]EC (2001).

eties.[1] Although the summary information gives an interesting snapshot of radioactive waste management practices, the document contains no information on procedures for clearing or exempting materials from regulatory control.

The committee's statement of work specifically requested a review of EU activities. The EU offers important comparisons with U.S. practices and regula-

---

[1]The INSC document is available on the Internet at <http://www2s.biglobe.ne.jp/~INSC/INSCAP/Radwaste.html>.

Basis for Clearance	Situation	Remarks
Implementation of Directive 96/29/ Euratom[b] by incorporation of existing regulations, except disposal of waste is expected in a few months		Basis for clearance is 10 µSv criterion Exemption orders exist that allow less restrictive clearance levels for naturally occurring radionuclides
January 19, 2001; memorandum from DOE Secretary (a) Metals recycle only within DOE (b) Moratorium and suspension remain (c) Environmental Impact Statement needed before regulations are revised (d) Reuse of lead and lead products	Pending the improved release criteria and information management recycle of scrap metals Pending USNRC decision to establish national volumetric standards	Other materials and equipment are released under DOE Order 5400.5,[d] which bases case-by-case approval on criteria of a small fraction of 1 µSv in a year and ALARA[e]
Table I of Regulatory Guide 1.86[f] for surface radioactivity	Ongoing USNRC study	Authorized release for disposal is possible on case-by-case basis

[d] DOE (1993a).
[e] ALARA = as low as reasonably achievable.
[f] AEC (1974).

SOURCE: USNRC (2001b).

tory trends, and it is a major trading partner of the United States for recycled materials, particularly metals.

CLEARANCE STANDARDS IN THE EUROPEAN UNION

Clearance practices in the EU are subject to a directive of the Council of the European Union, Directive Number 96/29/Euratom of May 13, 1996 (EU, 1996). The subject of this directive is "... laying down basic safety standards for the protection of the health of workers and the general public against the dangers arising from ionising radiation." Article 3, Section (2), defines the following exemptions to practices for the control of radioactive material if specified quantities or concentration limits are not exceeded (EU, 1996, p. 6):

No reporting need be required for practices involving the following:

(a) radioactive substances where the quantities involved do not exceed in total the exemption values set out in Column 2 of Table A to Annex I or in exceptional circumstances in an individual Member State different values authorized by the competent authorities that nevertheless satisfy the basic criteria set out in Annex I; or

(b) radioactive substances where the concentration[s] of activity per unit mass do not exceed the exemption values set out in Column 3 of Table A to Annex I or in

(c) exceptional circumstances in an individual Member State different values authorized by the competent authorities that nevertheless satisfy the basic criteria set out in Annex I; or

(d) ... [this item deals with sealed sources in devices that exceed the exemption limits but are devices that are approved by a Member State of the EU]; or

(e) ... [this item deals with electrical apparatus that can produce ionizing radiation]; or

(f) ... [this deals specifically with cathode ray tubes in x-ray equipment]; or

(g) material contaminated with radioactive substances resulting from authorized releases which competent authorities have declared not to be subject to further controls.

Table A to Annex I, which lists limits by nuclide, is reproduced in Appendix D of this report (see Table D-1). Annex I contains "Criteria to Be Considered for the Application of Article 3" in exempting a practice from regulatory control. For comparison, tables have been generated using the NUREG-1640 methodology discussed in Chapter 5 of this report assuming a dose level of 10 μSv/yr (1 mrem/yr) total effective dose equivalent (TEDE). These dose factors are given in Appendix D (see Table D-2) as information for the reader. The relationship between EU values, NUREG-1640 values, and other calculations of dose factors is discussed in Chapter 5.

The EU criterion of particular relevance to dose-based clearance standards is Paragraph 3, which allows member states to substitute their own limit values for those shown in Table A of Annex I, provided that both an individual dose limit and a condition on collective dose are met. The exact language of this "exemption" paragraph is included in Box 7-2.

BOX 7-2
Annex I (from Council Directive 96/29/EURATOM):
Criteria to Be Considered for the Application of Article 3

1. A practice may be exempted from the requirement to report without further consideration, in compliance with Article 3 (2) (a) or (b) respectively, if either the quantity or the activity concentration, as appropriate, of the relevant radionuclides does not exceed the values in column 2 or 3 of Table A.
2. The basic criteria for the calculation of the values in Table A, for the application of exemptions for practices, are as follows:
    (a) the radiological risks to individuals caused by the exempted practice are sufficiently low as to be of no regulatory concern; and
    (b) the collective radiological impact of the exempted practice is sufficiently low as to be of no regulatory concern under the prevailing circumstances; and
    (c) the exempted practice is inherently without radiological significance, with no appreciable likelihood of scenarios that could lead to a failure to meet the criteria in (a) and (b).
3. Exceptionally, as provided in Article 3, individual Member States may decide that a practice may be exempted where appropriate without further consideration, in accordance with the basic criteria, even if the relevant radionuclides deviate from the values in Table A, provided that the following criteria are met in all feasible circumstances:
    (a) the effective dose expected to be incurred by any member of the public due to the exempted practice is of the order of 10 μSv or less in a year; and
    (b) either the collective effective dose committed during one year of performance of the practice is no more than about 1 man-Sv or an assessment of the optimization of protection shows that exemption is the optimum option.
4. For radionuclides not listed in Table A, the competent authority shall assign appropriate values for the quantities and concentrations of activity per unit mass where the need arises. Values thus assigned shall be complementary to those in Table A.
5. The values laid down in Table A apply to the total inventory of radioactive substances held by a person or undertaking as part of a specific practice at any point in time.
6. Nuclides carrying the suffix "+" or "sec" in Table A represent parent nuclides in equilibrium with their correspondent daughter nuclides as listed in Table B. In this case the values given in Table A refer to the parent nuclide alone, but already take account of the daughter nuclide(s) present.
7. In all other cases of mixtures of more than one nuclide, the requirement for reporting may be waived if the sum of the ratios for each nuclide of the total amount present divided by the value listed in Table A is less than or equal to 1. This summation rule also applies to activity concentrations where the various nuclides concerned are contained in the same matrix.

SOURCE: EU (1996, Annex I).

In short, two mechanisms exist in the EU for clearing SRSM materials from regulatory control:

1. Materials can be released from regulatory control if the quantities and concentrations of activity per unit mass do not exceed the concentration limits listed in Column 3 of Table A in Annex I.
2. Competent regulatory authorities may use their own assessment process, conforming to the general approach used to derive the Table A values, to decide that a proposed exempting practice is within the principal individual and collective dose limits.

The EC has issued *Communication from the Commission concerning the implementation of Council Directive 96/29/Euratom* (EC, 1998a) describing how to implement Council Directive 96/29/Euratom (EU, 1996). With respect to Article 3 of the Directive 96/29/Euratom, the communication states:

> Article (3)(2) and Annex I specify the circumstances under which competent authorities may decide that reporting is not required. Member States are allowed to deviate from the values in Table A of Annex I in exceptional circumstances and subject to specified conditions.

This EC communication also contains information on how the values in Table A of Annex I were calculated:

> The exemption levels, which apply to practices, are worked out using scenarios, pathways and formulae presented in the report published by the Commission. (Radiation Protection No. 65, Principles and methods for establishing concentration and quantities [exemption values] below which reporting is not required in the European Directive, Luxembourg, 1993.)

A related EU directive on shipments of radioactive waste is officially titled *Council Directive 92/3/Euratom on the supervision and control of shipments of radioactive waste between Member States and into and out of the Community (OJ L 35, 12.2.92)* (EU, 1992). This directive controls the shipment of radioactive materials that have not been exempted or cleared from regulatory control. In addition, the Environmental Directorate of the EC has published the guidance document *Radiation Protection 89: Recommended Radiological Protection Criteria for the Recycling of Metals from the Dismantling of Nuclear Installations* (EC, 1998b), which provides activity standards for both surface and volume contamination of solid materials. These standards have been applied at several facilities in the EU. As indicated in Table 7-1, EU member nations are in various stages of developing detailed regulations to implement Directive 96/29/Euratom. There is a lack of uniformity of views regarding standards for materials that are candidates for release from further regulatory control, as described in the paper "Management of Slightly Contaminated Materials: Status and Issues" (Pescatore, 2001).

FINDINGS

Finding 7.1. The EU and the IAEA have each established a dose-based standard of 10 μSv/yr (1 mrem/yr) for the clearance of materials from regulatory control. A collective dose standard is also included, expressed as a committed dose equivalent of 1 man-Sv per year of exposure of the affected group (100 man-rem total effective dose equivalent per year).[2]

Finding 7.2. The EU has derived tables using a scenario assessment process against which radioactive solid materials can be evaluated for clearance.

Finding 7.3. A body of science, policy, and literature supports the development of the EU safety directives related to radioactive solid material clearance. In particular, the IAEA has developed policy guidance found in *Principles for the Exemption of Radiation Sources and Practices from Regulatory Control* (IAEA, 1988).

---

[2]Provisions exist in the EU safety directives for competent authorities in member states to develop alternative clearance guidance for special or specific circumstances.

8

Stakeholder Reactions and Involvement

PAST USNRC EFFORTS AT STAKEHOLDER INVOLVEMENT

This chapter reviews recent past and current efforts by the U.S. Nuclear Regulatory Commission (USNRC) to involve stakeholders in decision-making processes relevant to clearance standards for slightly radioactive solid material (SRSM). Three efforts by the USNRC to promote public involvement are particularly important and are discussed next: (1) the below regulatory concern (BRC) policy in the early 1990s, (2) the License Termination Rule (1992-1997), and (3) the 1999 issues paper that initiated a regulatory process for release of SRSM. The chapter then presents basic principles that the USNRC can follow to avoid past mistakes and involve its stakeholders more effectively.

The Below Regulatory Concern Effort

The BRC policy was intended to cover four basic clearance standards: (1) clearance of licensed facilities containing residual radioactivity after license termination; (2) distribution of consumer products containing small amounts of radioactivity; (3) disposal of solid wastes containing very low levels of radioactivity; and (4) recycling or reuse of solid materials containing very low levels of radioactivity (USNRC, 1991b). As noted in Chapter 2, promulgation of the BRC policy began in 1990 with a series of public meetings in which comments were obtained from various stakeholders (USNRC, 1991a). However, this public involvement process polarized as it progressed. Four of eight environmental and consumer groups that had been actively involved in the initial meetings refused to

136

discuss entering a consensus-building process because of the conditions set forth for entering into the process. These four organizations had been among the stakeholder groups most actively engaged in BRC issues.

Strong stakeholder opposition ultimately prompted the USNRC to defer action on petitions submitted by licensees for BRC exemptions (56 Federal Register 21631; May 10, 1991). The USNRC began an open, consensus-building process to clarify differences among affected parties and to work toward resolution of issues. However, the USNRC imposed two critical conditions on groups participating in the process. First, representatives from *all* parties who previously had a major interest in the BRC policy—as determined by the USNRC—were required to participate in a "core group." Second, all parties were required to agree to defer action on other avenues of relief (legislative, legal, or administrative) (USNRC, 1991d). These conditions and the general distrust engendered by the process resulted in continued boycott by certain groups. A letter from the Natural Resources Defense Council declining to participate was particularly persuasive in terminating the BRC process (NRDC, 1991).

The License Termination Rule

Following the withdrawal of the BRC policy, the USNRC decided to focus on issuing—in conjunction with an enhanced public participation process—a rule governing the clearance of facilities containing residual radioactivity (USNRC, 1992). The agency solicited input through a series of public workshops designed to identify issues, areas of concern, and disagreement. In addition, the normal notice and comment process was initiated. An initial draft rule was circulated by the USNRC staff along with public comments from the workshops on February 2, 1994 (59 Federal Register 4868). The additional comments on this initial rule were considered in the report to the Commission recommending a proposed rule for publication (USNRC, 1994). The Commission was to hear the final rule after another round of public comment in early May of 1995.

The report to the Commission gave special attention to three categories of comments critical of the initial draft proposed rule (USNRC, 1994):

- Comments that questioned the technical basis of the draft rule's 3 mrem/yr goal and 15 mrem/yr limit for individual dose;
- Suggestions by several licensees, industry groups, and the Environmental Protection Agency (EPA) that the 3 mrem/yr goal be dropped because it would become a de facto limit; and
- Comments indicating a need for greater guidance on demonstrating compliance with the rule's provisions.

The proposed rule (59 Federal Register 43200-43232; August 22, 1994) was published by the USNRC after considering the outcome of the workshops, the

National Environmental Policy Act (NEPA) scoping results, and the comments on the initial draft rule. The proposed rule dropped the 3 mrem/yr goal and retained the 15 mrem/yr standard. It retained the site-specific advisory boards that had been in the draft and endorsed the need for additional guidance on how the rule should be applied and enforced.

With publication of the proposed rule, the USNRC should have been able to conclude a successful public participation process. However, subsequent USNRC actions fundamentally undercut the consensus that had been achieved, further alienating many of those who had participated. The USNRC had announced an extension of the comment period for the proposed rule to January 20, 1995 (59 Federal Register 63733; December 9, 1994). Then on August 7, 1995, the USNRC announced (60 Federal Register 40117) an extension of the schedule for the final rule until early 1996 "to allow the NRC to more fully consider public comments received on the technical basis." That announcement of schedule extension noted the USNRC's intention to hold a public meeting in September 1995 to address specific issues and included the separate views of one commissioner questioning the adequacy of the technical basis for selecting a dose criterion of 15 mrem in contrast to 25 or 30 mrem. A letter dated September 25, 1995, from 10 environmental and consumer organizations objected to the "Commission's current move to hold a single workshop in Washington, D.C., to discuss a [new] proposal, . . . a possible 35 mr/yr clean-up standard [that] would substantially relax the final rule and is contrary to all of the remarks and comments [from the 1993 workshops]." The letter also charged that among other issues, "public comments in those sessions were a mandate for the most radiologically protective standard possible." The letter's authors asserted that the proposed rule was no longer adequate (Mariotte et al., 1995). Separately, the EPA objected to raising the standard from 15 mrem/yr to 25 or 30 mrem/yr because the higher limits would not adequately protect public health and the environment (EPA, 1997d).

Contrary to the consensus that had emerged from the extensive public process, the final rule (62 Federal Register 39058-39092; July 17, 1997) contained a 25 mrem/yr cleanup standard. It dropped the requirement for establishing site-specific advisory boards, substituting only broad performance criteria for obtaining such advice.

The 1999 Issues Paper and Current Stakeholder Involvement Efforts

The stated intention of the USNRC's June 1999 *Federal Register* notice (64 Federal Register 35090-35100; June 30, 1999) entitled "Release of Solid Materials at Licensed Facilities: Issues Paper, Scoping Process for Environmental Issues and Notice of Public Meetings" (the "1999 issues paper") was to initiate another "enhanced participatory process" for a proposed rule on clearance of SRSM (USNRC, 2000a). The 1999 issues paper established essentially three alternative actions (USNRC, 2000d, Attachment 1):

1. Do not conduct a rulemaking and proceed by continuing with current case-by-case practices.
2. Do not conduct a rulemaking and proceed by exploring options for updating existing guidance to improve consistency of criteria.
3. Conduct a rulemaking to develop a proposed rule.

If the third alternative, to proceed with rulemaking, were to be adopted, three technical approaches would be explored:

1. Permit release of solid material for unrestricted use if doses to the public from releases are less than a specified level.
2. Restrict release of materials to only certain authorized uses.
3. Prohibit release of material from areas where radioactive material has been stored—otherwise allow clearance.

As in previous efforts, the process centered around a series of public meetings. At these meetings, the USNRC once again asked environmental and consumer groups and other stakeholders to participate in a process that many of them had severely criticized and still doubted had been adequately reformed. This skepticism led some national environmental and consumer advocacy groups to boycott the public meetings intended to consider the issue of clearance of SRSM from USNRC-licensed facilities. Nevertheless, the USNRC received more than 900 comment letters.

The significant public concern expressed in these comments, combined with the boycott of the meetings, prompted the USNRC to hold an additional public meeting, conducted by the Commission and attended by representatives of a variety of stakeholder groups, including some that had boycotted earlier meetings (USNRC, 2000c).

Summaries of the stakeholder comments and oral statements from meetings on the 1999 issues paper are contained in a staff report to the USNRC (USNRC, 2000c) and in a consultant's report (USNRC, 2000d). Both sources provide the reader with some sense of the range of views expressed on the 1999 issues paper. However, they contain little detail about the number of comments in each category, the number of comments received that do not fall into the categories, or the intensity of the views expressed. The following analysis is the committee's attempt to fill in some of these details and fathom the extent and depth of reactions to the proposed alternatives. This analysis illustrates some key themes that the committee found in the diverse, sometimes conflicting, views of various stakeholders. It does not cover all of the groups that expressed opinions at the meetings, nor does it cover all possible opinions and options. For more detailed coverage, see Appendix F to this report, as well as the consultant's report on the meetings (USNRC, 2000d).

The positions expressed by stakeholder groups regarding the 1999 issues paper typically were similar to the positions articulated by the same groups during the BRC policy debate. Some of the strongly critical groups expressed views that certain policy options contained in the 1999 issues paper presented even greater risks than did the BRC policy. Their concern was that the Department of Energy (DOE), which they perceived as having large volumes of SRSM, was likely to handle that material in accordance with the USNRC approach to SRSM clearance.

Table 8-1 indicates the preferred alternatives for a number of stakeholder groups. The range of positions articulated is illustrated by the following list of stated views:

- Preclude any release of contaminated materials from regulatory control.
- Continue the USNRC's case-by-case process.
- Promulgate a conditional clearance standard (e.g., landfill disposal).
- Promulgate a clearance standard.
- Delay decision until a process is established for arriving at a consensus.

The alternatives presented in the issues paper—represented in Table 8-1 by the "Do Not Conduct a Rulemaking" and "Conduct a Rulemaking" columns—do not capture the full spectrum of alternatives favored by stakeholder groups. For example, many of the environmental and consumer groups that expressed an opinion criticized the USNRC for failure to include a "no release" alternative (see columns for "Other" in Table 8-1).

Three major themes emerged from the committee's analysis of the complete range of stakeholder views expressed in response to the 1999 issues paper:

- *Theme 1. There is little support from stakeholder groups for a clearance standard for SRSM.* Although agreement states and the nuclear industry favor some form of clearance standard, many consumer and environmental groups and certain affected industry organizations do not. Environmental groups expressed concern about risks to human health from clearance. The metals and concrete industries expressed concern that the presence of radioactive materials in their products would negatively affect their sales due to public fear. The metals industry also feared an economic impact if public confidence were decreased in the safety of steel products.
- *Theme 2. There is a legacy of institutional distrust of the USNRC by some of its stakeholder groups, particularly the environmental and consumer advocacy groups.* The three regulatory events described above have contributed to this distrust of USNRC by certain stakeholder groups. Other reasons, based on stakeholders' perceptions that may or may not have a basis in fact, are evident in the public comments received by the USNRC

and this committee. Among the factors that undermine trust are the following:

— The USNRC and the DOE are perceived as not having fully disclosed the risks and uncertainties associated with establishing a clearance standard.
— The perception is that the true purpose of establishing a clearance standard is to provide regulatory cover for DOE's efforts to recycle radioactive materials.[1]
— The USNRC is perceived to have focused almost exclusively on economic benefits rather than protecting human health and the environment.
— The perception that the USNRC lacks the capacity to regulate the implementation of a clearance standard effectively.
— The perception that USNRC's public participation process is implemented mechanically, with little or no commitment to comprehending and addressing stakeholder concerns.

Appendix F illustrates several of these perceptions in depth, categorizing them with respect to the views of particular groups.

- *Theme 3. Numerous stakeholders are unclear about the meaning or import of certain technical terms and issues.* Among the sources of confusion are the panoply of radiation control units of measure (e.g., sieverts, rems, becquerels) and technical distinctions such as those between surface contamination and volume (or volumetric) contamination, between (unconditional) clearance and conditional clearance, or between exclusions and exemptions.

Summary of Stakeholder Views

In summary, the committee's review of the record on the BRC policy, the License Termination Rule, and the 1999 issues paper found that many stakeholders distrust the USNRC and remain confused about important technical questions. There are misperceptions about intentions on both sides, and the USNRC has not been effective in its risk communication. There is also no consensus evident among stakeholder groups about the options for regulating disposal of SRSM. The USNRC must overcome serious levels of distrust, generated by its actions during the BRC policy and License Termination Rule efforts, before the

---

[1] For a brief account of circumstances cited by some groups to support this strongly negative perception, see the section below on "DOE Recycling of SRSM: The Oak Ridge Project."

TABLE 8-1 Matrix of Stakeholder Perspectives

Stakeholder	Do Not Conduct a Rulemaking — Continue Case by Case	Conduct a Rulemaking — Release for Unrestricted Use (Clearance)
Nuclear Information and Resource Service		
Public Citizen		
New England Coalition on Nuclear Pollution		
Allied Industrial Chemical and Energy Workers Union		
Natural Resources Defense Council		
Steel Manufacturers Association		
American Iron and Steel Institute		
National Ready- Mixed Concrete Association		
Metals Industry Recycling Coalition		
Association of Radioactive Metal Recyclers		
Association of State and Territorial Solid Waste Management Officials		
Illinois Department of Nuclear Safety, representing 49 States[b]	X[c]	
Health Physics Society		X[d]
American Nuclear Society		X[d]
Nuclear Energy Institute		X[d]
Conference of Radiation Control Program Directors		X[d]

[a]Authorized use includes both licensed (nuclear) use and unlicensed use (landfills, bridge supports).

[b]More specifically, representing the Conference of Radiation Control Program Directors and the Organization of Agreement States.

[c]These groups want to continue case by case but with uniform national criteria.

[d]Group expressed view that some special exceptions might apply, i.e., for metals industry.

Restrict Release to Certain Authorized Uses (Conditional Clearance)[a]	Other		
	No Release (No Option Specified, But Want the Solid Waste Isolated from General Commerce)	Cannot Engage in Dialogue Because Dialogue Process Is Tainted	Recommend Delaying Decision on a Rule Until Stakeholder Views Are Integrated with USNRC Decision Framework
	X		
	X	X	
	X		
		X	
			X
X			
X			
X			
X			
X			X
			X

expanded public participation process associated with the 1999 issues paper is likely to succeed.

RISK COMMUNICATION AND ITS ROLE IN THE RULEMAKING PROCESS

Approaches for effective risk communication have become highly sophisticated over the past 10 to 15 years. A study committee of the National Research Council has defined risk communication as follows (NRC, 1989, p. 21):

> [A]n interactive process of exchange of information and opinion among individuals, groups and institutions. It involves multiple messages about the nature of risk and other messages, not strictly about risk, that express concerns, opinions, or reactions to risk messages or to legal and institutional arrangements for risk management.

According to this definition, risk communication is a reciprocal process, not an attempt by an agency to "sell" its program to the public. If decisions are not negotiable, then the agency should not waste stakeholders' time (Omenn, 1997, p. 18). The approach must embody the principle, articulated by Thomas Jefferson, that there is "no safe depository of the ultimate powers of society but the people themselves; and if we think them not enlightened enough to exercise their control with a wholesome discretion, the remedy is not to take it from them, but to inform their discretion" (Jefferson, 1820). Risk communication succeeds when it promotes a deeper understanding of the issues and satisfies individuals involved that they are adequately informed within the limits of available knowledge and that their views have been fairly considered.

Further, the concept of risk communication is consistent with federal laws on open government, which were meant to promote public participation in agency decision making. Among these laws are the Administrative Procedures Act, the Federal Advisory Committee Act, the Government in the Sunshine Act, the National Environmental Policy Act, and the Freedom of Information Act.

Communicating the risks and benefits of a clearance standard to the public is challenging because of both the fears associated with radiation and the technical nature of the issues. The USNRC has successfully engaged in risk communication in limited contexts, such as the initial public participation process during development of the License Termination Rule. The USNRC's inability to follow through on the 1994 consensus is an equally compelling example of poor risk management and communication. The results of these errors and others during the BRC policy effort have included a stalemate on SRSM clearance and disposal issues, as well as increased distrust of the USNRC.

The USNRC through a series of studies it commissioned and finished in 1999, has been made fully aware of the "state of the art" in using risk communication with both the public and decision makers. If the USNRC implements the information contained in the reports, their efforts will be better informed than past work that employed, but did not follow through with, participatory processes and risk communication. Interestingly, the commissioned studies view sharing power and empowering the public in decision-making processes as a critical function of risk communication with the public and a crucial step in building trust or credibility, deeming it the "ultimate solution to situations of [existing] distrust" (Bier, 1999a, 1999b).

Stakeholders' Distrust and Deficiencies in the USNRC Process

The USNRC's request for stakeholder input should, in principle, be acceptable as an honest effort to respect and consider all stakeholder views. For a variety of reasons discussed above, many stakeholder groups do not view it this way. Many of the stakeholder groups that boycotted the initial workshops on the most recent reconsideration of the SRSM issue expressed skepticism that the USNRC was substantively considering and responding to their views and expressed concern that USNRC had not solicited their input prior to publishing the 1999 issues paper. These concerns are not directed toward scientific or technical issues but to issues of *process*.

The USNRC maintains the final responsibility for any rule or change in policy, but within its statutory limitations there is a great deal of latitude for involving stakeholders. Legitimacy can be achieved only through fostering trust in the agency's integrity, fairness, honesty, and competence (Pijawka and Mushkatel, 1992). If the process appears to be biased, if the communications are one-sided and technically obscure, or if uncertainties are disregarded, many stakeholders will view both the process and the outcome as illegitimate. When this occurs, groups seek other avenues, such as the courts or Congress, through which to be heard.

DOE Recycling of SRSM: The Oak Ridge Project

Stakeholders' concerns on the clearance issue have been influenced by their experience with DOE projects, as well as by their experience directly with the USNRC. A DOE plan to recycle approximately 100,000 tons of nickel and steel removed from the K-25 gaseous diffusion plant at Oak Ridge, Tennessee, resulted in the erosion of stakeholder trust. DOE proposed to remove and recycle the metals without completing an environmental impact statement—despite the size and novelty of the project. Moreover, a report by a National Research Council committee had previously recommended that public participation and support were critical to any such effort (NRC, 1996). Yet, the DOE project was initiated with essentially no public review or involvement, and regulatory approval to clear radioactively contaminated materials through an agreement state-licensed facility was conducted with no public process.

Environmental groups' concerns were confirmed when a contractor for the project, BNFL, was found to have an inadequate training program for employees, a deficient procurement system, problems in laboratory quality control, and several important violations of Occupational Safety and Health Administration standards. The DOE Inspector General confirmed these problems in a September 2000 report (DOE, 2000). The Inspector General also found that BNFL's surveys of contaminated materials were not conducted accurately, that employees were not adequately supervised, and that these problems posed an increased risk to the public (DOE, 2000, p. 2).

One of BNFL's partners on the Oak Ridge project was Science Applications International Corporation (SAIC), with whom the USNRC had also contracted to perform the technical analysis for NUREG-1640. In November 1999, the Paper, Allied-Industrial, Chemical, and Energy Workers International Union, which represents hourly workers at Oak Ridge National Laboratory, charged that the SAIC contract violated federal conflict-of-interest regulations precluding contractors from conducting work for the government that could benefit a private-sector client. In December 1999 the USNRC issued a stop-work order to SAIC, and in March 2000 it terminated the SAIC contract.

In July 2001, DOE announced plans to perform a Programmatic Environmental Impact Statement on scrap metal disposition, recycling, and clearance across its complex. SAIC was the contractor initially selected to undertake this work (Inside NRC, 2001). DOE canceled the SAIC contract on July 25, 2001, after environmental groups and an influential member of Congress raised concerns about possible bias stemming from SAIC's earlier involvement as a subcontractor to BNFL in the nickel recycling project (Zuckerbrod, 2001).

Hence, DOE's approach to the K-25 metals recycling, the subsequent problems with one DOE contractor for that project, and the links between that contractor and a second DOE contractor have further undermined the USNRC's credibility with some stakeholders. These stakeholders suggest that the two agencies are collaborating behind the scenes (i.e., "conspiring") to establish standards allowing clearance of SRSM.

The Importance of Trust

In the literature on public involvement, institutional trust is widely viewed as the single most important factor influencing the acceptance of controversial government policies (Raynor and Cantor, 1987; Flynn et al., 1992; Pijawka and Mushkatel, 1992). Trust is often characterized as a collection of attributes, such as honesty, fairness, integrity, competence, and consistency (DOE, 1993b). Research studies indicate that individuals accept higher levels of risk, or perceive risk as being lower, if they trust the agency setting the policy. The agency, however, must be perceived as honestly presenting the level of risk associated with the policy and as having the competence to evaluate the risks.

When an agency does not address issues consistently or is shown to have misinformed the public, stakeholder mistrust develops. By contrast, the more transparent a decision-making process is, the more likely are stakeholders to perceive the agency as having nothing to hide. The USNRC has lost the trust and confidence of some of its important stakeholder groups. It now must either work to regain their trust or continue to contend with an increasingly adversarial relationship. Some encouragement can be gleaned from studies showing that although trust is easy to lose and difficult to regain, it can be rebuilt through a concerted and sustained effort (Kasperson et al., 1988). The USNRC will have regained trust when it has significant participation by a broad base of stakeholders in its rulemaking process.

STAKEHOLDER INVOLVEMENT: METHODS AND SUCCESSES

The USNRC has had limited success in obtaining meaningful stakeholder involvement. Even so, determining the proper strategy or process to increase effective public participation and rebuild the trust of stakeholder groups will be difficult. Various types of dispute resolution techniques that may be appropriate at steps along the way include unassisted procedures or third-party assistance, including facilitation, mediation, fact finding, and nonbinding arbitration. Some authorities have found partnering techniques to be successful in avoiding disputes (Creighton and Priscoli, 1996).

Formalized public involvement, such as the workshops that the USNRC has conducted recently, is designed to give stakeholders an opportunity to be heard prior to a decision and to involve them in the framing of problems and solutions. Approaches such as facilitation, fact finding, mediation, and nonbinding arbitration allow stakeholders to participate in the evaluation of alternatives, impacts, and proposed decisions (see Figure 8-1). Some forms of dispute resolution are designed to require stakeholders' approval before a final decision is made (Creighton and Priscoli, 1996).

Some determinations must be made before selecting and moving forward with any of these methods or techniques for public participation. In particular, it is critically important that the agency and the stakeholders both believe that they can benefit from the process whether it is a public consensus-building process or an alternative dispute resolution approach. That is, the entities must believe that the outcome is more likely to be favorable to them if they participate in the joint process rather than remain outside it.[2] When this belief is lacking on either side, these processes are not appropriate. If the agency is bound legally to one option or if the agency does not believe that stakeholder involvement is important and worthwhile, these methods should not be employed.

---

[2]Janesse Brewer, the Keystone Center, presentation to the committee, June 27, 2001.

UNASSISTED PROCEDURES	Information Exchange Meetings	Interest-Based Negotiation			

Increased Structure/Formality →

THIRD-PARTY ASSISTANCE	Facilitation	Mediation	Fact-Finding	Disputes Review Board	Non-Binding Arbitration

Process ————————→ Substance
Form of Third-Party Assistance

THIRD-PARTY DECISION MAKING	Binding Arbitration	Administrative Hearings	Litigation

Increased Structure/Time →

DISPUTE PREVENTION	Partnering

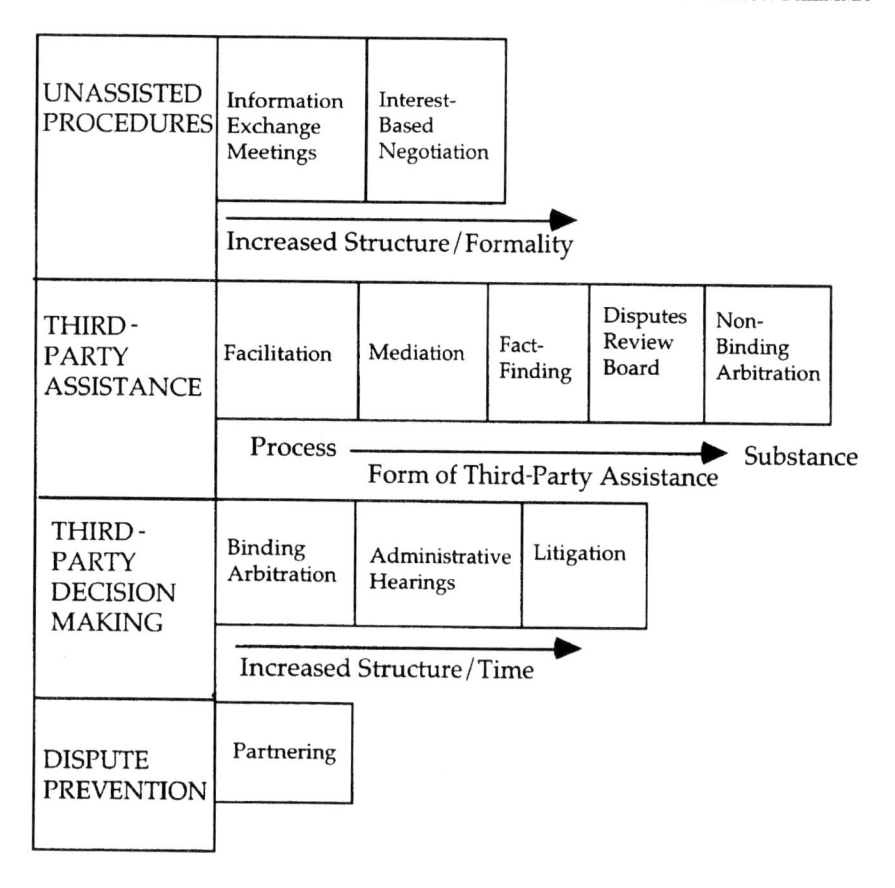

FIGURE 8-1 Dispute resolution techniques. SOURCE: Creighton and Priscoli (1996, p. 24).

An agency can gain several benefits from using public involvement strategies appropriately. These benefits include not only building legitimacy for decisions but also gaining new information and perspectives. The affected public may gain new information and perspectives as well, and the process can keep all constituencies better informed. However, if parties on either side are not acting in good faith, such methods may do more harm than good.

Both the EPA and the U.S. Army Corps of Engineers (USACE) have extensive alternative dispute resolution programs that have received widespread attention (Creighton and Priscoli, 1996). The EPA has published for review a draft plan for public involvement (EPA, 2000). The U.S. Army has successfully used a dialogue process designed by the Keystone Center to gain public acceptance of an alternative technology for the destruction of chemical weapons. The USACE and the Department of Defense are using partnering approaches extensively to minimize disputes. The DOE, which has used site-specific advisory boards extensively, has recently retained a public involvement consulting firm (Creighton and Creighton) to design materials for its public involvement processes. The U.S. Bureau of Reclamation has conducted an extensive review of its public involvement programs and is revising them.

No single approach is best for all situations or for all agencies. Much depends on an agency's true goals. If the USNRC truly believes that it is *important and worthwhile* to involve stakeholders, then it should assess the willingness of stakeholder groups to begin a dialogue. This dialogue will have to address not only items contained in the 1999 issues paper but also issues that some stakeholder groups claim have been omitted. The assessment should address stakeholder views about desirable and feasible mechanisms for obtaining sustained stakeholder input into (1) how issues should be framed and (2) how decision processes can be made transparent and open. This assessment should be viewed as *just the first step* toward rebuilding the credibility of the agency and beginning to reestablish trust by stakeholders. In addition, it is critical that the dialogue clearly spells out *up front* what flexibility the USNRC has in responding to specific stakeholder concerns and where it feels it is statutorily precluded from taking action. This delineation of where action is feasible will allow stakeholders both to know they can have some influence and to determine if this amount of influence on the outcome is sufficient to justify their participation in the process. In order to increase the belief of stakeholder groups that their input matters, it is vital that the USNRC provide ongoing feedback as to how the agency is utilizing the input from the dialogue group. Feedback should include both the identification of when and how input affected decisions and the reasons input did not have an effect.

The USNRC, like many other federal agencies, has tended to rely on a small and closed circle of contractors for certain services. Although a tight circle of support contractors may simplify procurement of specialized technical services, it fosters negative perceptions, by those outside the circle, of the openness and fairness of the process. These perceptions often underlie and reinforce beliefs that USNRC contractors are not adequately trained, have not exercised credible efforts to meet safety and quality standards, and often represent too closely the interests and perspectives of the regulated industry.

As noted, other agencies have adopted innovative and far-reaching approaches to public involvement, alternative dispute resolution, and consensus building. The USNRC should reach out to the contractors that have been involved in these programs for the EPA, the Army (including the USACE[3] and the chemical weapons demilitarization programs), and other agencies.

---

[3]The alternative dispute resolution handbook developed for the USACE (Creighton and Priscoli, 1996).

FINDINGS

Finding 8.1. The USNRC involved stakeholders in the processes for the BRC policy and the License Termination Rule for decommissioning, as well as in the initial stages of considering standards for release of SRSM. Despite these efforts, environmental and consumer advocacy groups remain concerned with radiation effects, and industrial groups continue to be concerned with the potential economic consequences of the clearance of SRSM.

Finding 8.2. Most of the issues of concern to those stakeholder groups that oppose the USNRC's recent efforts to establish a rule for the release of SRSM are the same issues expressed by these groups 10 years ago during the effort to establish the BRC policy. The committee's review of the record on the BRC policy, the License Termination Rule, and the 1999 issues paper found that stakeholders distrust the USNRC and remain confused about important technical questions. There are misperceptions about intentions on both sides, and the USNRC has not been effective in its risk communication.

Finding 8.3. Stakeholder groups differed in their viewpoints on regulating disposition of SRSM. Generally, professional societies associated with the nuclear industry supported clearance, industrial groups endorsed conditional clearance, and environmental groups opposed any type of clearance. However, much of the opposition to a clearance standard was associated with recycling metal SRSM into general commerce.

Finding 8.4. A legacy of distrust of the USNRC has developed among most of the environmental stakeholder groups. This distrust results from their experience with the BRC policy, the License Termination Rule, and the 1999 issues paper on the release of SRSM. Reestablishing trust will require concerted and sustained effort by the USNRC, premised on a belief that stakeholder involvement will be important and worthwhile, as well as a prerequisite for making progress.

9

A Framework and Process for Decision Making

PROBLEMS WITH THE CURRENT APPROACH

The current approach for releasing radioactive materials from facilities licensed by the U.S. Nuclear Regulatory Commission (USNRC) is based on Regulatory Guide 1.86 (AEC, 1974), USNRC guidance memoranda, and the case-by-case application of section 2002 of 10 CFR Part 20 by USNRC and its agreement states. Several problems with this approach were pointed out in presentations to the study committee (see details in Chapters 2 and 8). From an administrative perspective, the major concerns expressed were that this approach does not handle volume contamination generically and that the case-by-case approach may lead to inconsistent determinations from one case to another. Another point made was that this approach and the acceptable surface contamination levels in Table I of Regulatory Guide 1.86 are 27 years old; they have not kept up with international developments of release standards, many of which are risk based (see Chapter 7). Also, the regulatory guidance was not adopted through rulemaking and hence was not submitted for public comment or review. The case-by-case applications for release produce additional workload and costs for the USNRC, but this burden appears manageable for the foreseeable future.

From the licensees' perspective, the major concerns expressed to the committee were that this approach is unpredictable and costly, and creates undesirable operational impacts. Licensees also expressed concern about future liabilities if materials released under Regulatory Guide 1.86 are later suspected to have caused harm.

151

From the perspective of environmental groups and some members of the public, a major concern with the current case-by-case approach is that it allows unrestricted uses of slightly radioactive solid material (SRSM) once it clears the surface contamination limits. However, representatives of this perspective typically do not favor dose-based standards as a remedy; they prefer a no-release approach. In addition, environmental groups criticized the current approach as being largely administrative and precluding the possibility of public scrutiny or external review.

For the above reasons and more, various stakeholders, including licensees, and other interested parties have argued for modifying or replacing the current approach. Their proposals for an alternative approach differ widely, ranging from a strict no-release policy favored by some to a dose-based standard for unconditional release favored by others. Given these different and strongly held views, the development, evaluation, and implementation of a regulatory approach will likely create substantial controversy and debate. It will take significant time and effort to develop an acceptable solution.

The committee recognizes that there are problems with the current approach and that a new approach is needed for many of the reasons stated by the stakeholders. However, the committee has not found any evidence that the problems with the current approach cause significant health effects or amount to an immediate crisis. The committee therefore concludes that it is possible for the USNRC to conduct, with deliberate speed, a thorough analysis and evaluation of several alternative approaches to the disposition of SRSM including a broad-based stakeholder involvement process.

THE DECISION-MAKING PROCESS

The USNRC has two important choices when considering a decision on the disposition of SRSM. The first choice is what kind of decision process to use—for example, a regular rulemaking process or an enhanced participatory process. The second choice is which alternatives for the disposition of SRSM it should study and evaluate. This section discusses process options. The next section describes a systematic framework for developing, analyzing, and evaluating disposition alternatives within this process.

The USNRC has various process options for making the decision about the disposition of SRSM. One possibility is to follow a variation of the National Environmental Policy Act (NEPA) process. NEPA provides a widely accepted structure for the announcement of a proposal by an agency, for solicitation of public input as to the appropriate range of alternatives and impacts to analyze through a scoping process, and for subsequent review of environmental analyses with public input. In addition, the NEPA concept of tiering will allow the USNRC to obtain input on issues of broad scope first and later move to NEPA review of increasingly specific options.

The USNRC used a scoping NEPA process in parallel with its enhanced participatory rulemaking process during 1992-1997, while developing its License Termination Rule, 10 CFR Part 20, Subpart E. The USNRC might reconsider that experience, and the experience with the below regulatory concern (BRC) policy statement that preceded it, to evaluate a tiered NEPA approach overall. The BRC process did not use an enhanced open approach and had severe difficulties. The enhanced participatory process for the License Termination Rule was an open NEPA approach and appeared to have achieved consensus until the USNRC's process changed, following the issuance of the proposed rule.

As explained in Chapter 2, the BRC policy statement was required of USNRC in response to Section 10 of the Low-Level Radioactive Waste Policy Amendments Act of 1985 (LLWPAA; 42 U.S.C. §2021j), which was specifically directed at defining a release standard for radioactive material that was at such a low level that it would be "below regulatory concern." The BRC policy statement addressed this statutory provision with an overarching dose-based or risk-based policy. The policy would have provided guidance for setting BRC standards for radioactive waste, residues at license termination, exemption of radioactivity in consumer products, and general release of materials for recycle or reuse.

If a tiered NEPA process had been followed, the USNRC might have begun by developing a draft policy statement, with full public input and participation. Then it would have proceeded with separate NEPA processes for each of the subsequent decisions. Instead of this tiered NEPA process, the USNRC developed and published the BRC policy statement in 1990 but turned to public consensus building only after receiving severe negative reactions to the policy. Public acceptance was not built step by step, nor was the policy developed in an iterative manner. The consensus process failed, and the BRC policy was first put on hold (56 Federal Register 36068; July 30, 1991) and then rescinded (58 Federal Register 44610; August 24, 1993). Since that failure to establish a broad policy, the lack of a top tier—an overarching policy—appears to have significantly hindered progress with the subsequent License Termination Rule and the development of standards for release of SRSM.

The USNRC decision processes can be improved by including a broad range of affected groups and individuals. Administrative appeals processes and administrative guidelines may have to be altered to ensure greater access to the USNRC's decision-making process by a broader range of affected individuals, industries, and interested parties. The goal should be to develop a process that solicits input broadly, while remaining flexible, open, transparent, and fair.

In addition, compared to some of the more recent national health and safety legislation (such as the Resource Conservation and Recovery Act [RCRA]; the Comprehensive Environmental Response Compensation and Liability Act [CERCLA]; the Clean Air Act, and the Clean Water Act) the USNRC's fundamental legislation, the Atomic Energy Act (AEA) provides a somewhat less extensive legal basis for citizens' suit challenges or public review. However, the legal basis is fully adequate if used properly. Whatever the AEA's shortcomings might be in this regard, the USNRC can and must employ the appropriate mechanisms to reach out to develop stakeholder participation, acceptance, and (eventually) support.

It is vital that any decision process for developing policies on clearance of SRSM begins from a broad set of alternatives. Among the alternatives could be options beyond just clearance of materials from licensed sites. In particular, the committee believes that it would be useful to consider alternatives beyond a clearance standard by looking at issues concerning the broader range of low-activity radioactive materials. For example, a broad-based scoping process could also include consideration of whether the USNRC should regulate naturally occurring and accelerator-produced radioactive material (NARM) and naturally occurring radioactive material (NORM) by some national standards rather than continuing with state-only regulation of these categories of radioactive materials.

The USNRC might consider supplementing its decision process with enhanced and expanded use of public advisory committees. Many federal agencies include members of the broader public—not just highly technical experts—on their advisory committees. The result of using NEPA, a broad scoping process, more iterative development of proposals, and broader participation on advisory committees would be greater and broader public participation in the USNRC decision-making process.

As the regulatory body, the Commission holds the statutory decision-making authority. Some concerned groups perceive the Commission and USNRC staff as nonresponsive to public input. In addition, many observers perceive the Commission and staff as not operating cohesively. Unless confidence and trust in the USNRC increase, acceptance by the public and Congress of a clearance or conditional clearance standard is unlikely.

Any process to develop a release standard might be enhanced by using professional facilitators. During the BRC process, the Commission called on one of the USNRC staff to lead the attempt at building consensus for BRC. The staff then recruited a professional facilitator, who worked on BRC and other matters. For the enhanced participatory rulemaking effort, the USNRC engaged the services of the Keystone Center, a group of professional facilitators. In the long run, the USNRC might benefit from further pursuit of facilitated participation processes to increase the likelihood of productive public involvement.

A SYSTEMATIC DECISION FRAMEWORK

Several alternatives exist for the disposition of SRSM: the current case-by-case approach, a no-release (from regulatory control) alternative, clearance, and conditional clearance. In addition, there are many combinations, types, and levels of possible standards and several possible clearance conditions worth considering. Impacts to be considered include public health, costs and benefits, consistency with existing national and international analysis and regulations, and public perceptions and acceptance. This section first defines a logical set of alternatives for disposition of SRSM, ending with the finding that for practical purposes, only a few alternatives merit further consideration. It then develops a list of impacts that should be examined when evaluating these alternatives.

Alternatives

In its statement of work (see Appendix C), the study committee was asked to consider the following alternatives for the disposition of SRSM from USNRC-licensed facilities:

1. Continue the current system of case-by-case decision;
2. Establish a national standard by rulemaking or other approaches; and
3. Consider other alternative approaches.

After gathering information and deliberating on the range of possible approaches, the committee decided to address two "other" approaches in some detail:

1. A no-release policy, and
2. Establishment of a national standard with conditions on the uses of released materials.

At the general level, there are thus four policy alternatives to address:

1. Case-by-case approach (the USNRC or an agreement state approves specific license conditions in accordance with Regulatory Guide 1.86 or modifications);
2. Clearance standard (unrestricted release of materials that meet the standard);
3. Conditional clearance standard (restricted release of materials that meet the standard); and
4. No releases of licensed material.

There are many possible variants for some of these alternatives. Box 9-1 illustrates some of these variants.

Not all of the alternatives in Box 9-1 merit detailed consideration here. For example, the committee found little support for minor modifications of the current approach. One such modification would be to develop additional criteria for volume contamination, based on a dose assessment, and apply these criteria on a case-by-case basis. As a second example, stakeholders who prefer a national

> **BOX 9-1**
> **Policy Alternatives for Releasing Slightly Radioactive Solid Materials**
>
> *Case-by-Case Approach*
>
> - Current approach: USNRC or agreement state approves specific license conditions
> - Additional criteria for volume contamination
> - Restrictions on reuse (see examples below, under conditional clearance)
>
> *Clearance Standard*
>
> - Dose based (based on risk to an individual or population caused by exposure to radiation)
> - Source based (based on surface or volume radioactivity concentration of the contaminated solid material)
>
> *Conditional Clearance Standard*
>
> - Dose based (based on risk to an individual or population caused by exposure to radiation)
>   - Beneficial reuse in controlled environments (e.g., metal for shield blocks in USNRC licensed or Department of Energy [DOE] facilities)
>   - Limited reuse for low-exposure scenarios (e.g., concrete rubble base for roads)
>   - Landfill disposal
> - Source based (based on surface or volume radioactivity concentration of the contaminated solid material)
>   - Beneficial reuse in controlled environments (e.g., metal for shield blocks in USNRC licensed or DOE facilities)
>   - Limited reuse for low-exposure scenarios (e.g., concrete rubble base for roads)
>   - Landfill disposal
>
> *No Release*
>
> - All slightly radioactive solid materials are disposed of at licensed LLRW sites.

standard (for unconditional or conditional clearance) typically argue for a dose-based standard rather than a source-based standard. Therefore, source-based variants for clearance standards are not addressed further herein.

Based on these and similar observations from its information gathering efforts, the committee focused on the following six policy alternatives and variants:

1. Case-by-case approach (pursuant to Section 2002 of 10 CFR Part 20 or possible modifications);
2. Dose-based clearance standard (unrestricted reuse, including commercial recycling);
3. Dose-based conditional clearance standard (beneficial reuse in controlled environments, e.g., shield blocks at Department of Energy [DOE] facilities);
4. Dose-based conditional clearance standard (commercial reuse for low-exposure scenarios, e.g., concrete rubble base for roads);
5. Dose-based conditional clearance standard (landfill disposal); and
6. No release (all SRSM is disposed of at licensed low-level radioactive waste [LLRW] sites).

The current case-by-case approach can be improved by developing additional criteria for volume contamination, possibly based on a dose assessment, using coefficients similar to those currently under development for the draft NUREG-1640.

Several possible dose limits for use in a dose-based standard have been discussed, including annual doses of 1 µSv (0.1 mrem), 10 µSv (1 mrem), or 100 µSv (10 mrem). Placing conditions on clearance has the effect of limiting the potential exposure scenarios. For example, suppose SRSM is cleared under a dose-based standard of 10 µSv/yr (1 mrem/yr) for landfill disposal only. If the same secondary activity standard were kept, the maximum individual dose would be lowered for most radionuclides, because the highest doses without the landfill restrictions apply to transport and factory workers, who would no longer be exposed on the job.[1] On the other hand, if the secondary activity standard is adjusted upward under a landfill restriction to allow the primary dose standard to be reached in the new critical group, then it would be possible to release SRSM with higher concentration under a conditional clearance standard than it would under an (unconditional) clearance standard.

The following discussion provides a few examples of the range and type of policy alternatives that the committee recommends to the USNRC. It may even be reasonable to consider alternative dose standards for different conditional clearance conditions. For example, if the restriction is beneficial reuse in controlled environments, a dose standard of 100 µSv/yr (10 mrem/yr) may be reasonable since exposure limits for nuclear workers are typically much higher (50,000 µSv/yr or 5,000 mrem/yr).

---

[1] The committee notes that modeling of exposed groups in draft NUREG-1640 (USNRC, 1998b) specifically rules out residential use of postclosure property. Had such a restriction not been made, landfills would become the critical group for some radionuclides and hence would already represent the maximum dose for these radionuclides.

Impacts of Alternative Regulatory Approaches

Many participants in the study committee's information-gathering meetings expressed concerns, issues, preferred outcomes, and objections in response to some of the alternatives discussed above. As discussed in Chapter 8, the USNRC has not gained widespread public trust in its recent rulemakings. For example, environmental groups objected to any standard that allowed the release of SRSM into commerce. They argued that this would create an unnecessary health risk with unknown cumulative effects. Some licensee representatives expressed concerns about liability risks and economic costs of regulation. Representatives from the steel and concrete industries worried about the possible stigmatization of their products if it became known that some of their materials might include radioactive contamination, no matter how slight.

The committee drew on these comments, together with the numerous statements of issues and concerns submitted in response to USNRC's June 1999 *Federal Register* notice (64 Federal Register 35090-35100; June 30, 1999) entitled "Release of Solid Materials at Licensed Facilities: Issues Paper, Scoping Process for Environmental Issues and Notice of Public Meetings" and public hearings in the fall of 1999 (see Appendix F), to create a generic list of impacts for consideration when evaluating alternatives for disposing of SRSM. This list is shown in Box 9-2 and discussed below.

Health Impacts

The primary objective of any alternative for the disposition of SRSM is to ensure that there are minimal health impacts for any individual and the public at large. Much of the work on dose-based standards (e.g., draft NUREG-1640) has focused on specific scenarios for individuals with the potentially highest doses from released materials. However, the committee also heard concerns about the potential for multiple exposures and collective doses, especially cumulative doses from multiple commercial products containing SRSM. To address these concerns, risk assessments must consider not only maximally exposed individuals and direct health impacts from a single source, but also the potentially exposed population and cumulative impacts from multiple sources.

There may also be indirect and unintended impacts from implementing alternative approaches. For example, under the current approach, radioactive materials must be shipped over long distances, usually by truck. One waste broker (Duratek, Inc.) estimated that its trucks drive about 6 million miles per year. With increased decommissioning activities, these shipment miles will increase substantially, thus increasing the probability of accidents, however low the probability per mile might be.

The conditional clearance option, by allowing disposal of SRSM in Subtitle C or D landfills, would reduce both the transportation miles and the associated

BOX 9-2
Possible Impacts of Alternatives for Slightly Radioactive Solid Materials from USNRC-Licensed Facilities

Health Impacts

- Dose to maximum exposed individuals
- Collective dose
- Cumulative impacts from multiple exposures
- Indirect and unintended health impacts

Environmental Impacts

- Transportation
- Disposal

Direct Costs

- Licensee waste management cost
- USNRC regulatory cost
- Other agencies' regulatory cost

Indirect Costs

- Licensee potential liabilities
- Product stigmatization (steel and concrete)

Direct Benefits

- Licensee benefits from resale of materials
- Reduction of operational expenses

Consistency with Existing Regulations

- International
- National (U.S. Environmental Protection Agency, other USNRC regulations)
- State and local

Implementation and Enforcement

- Ability to track the chain of custody of released materials
- Ability to detect violations
- Ability to enforce sanctions for violations
- Ability to detect, measure and monitor low levels of radioactivity

Public Perception

- Public trust and acceptability
- Public fears and concerns

transportation risks. The greater number of such landfills in the United States, relative to the three LLRW disposal facilities, means a much greater likelihood of a landfill being in close proximity to the power reactor that is undergoing decommissioning.

Environmental Impacts

Alternative approaches to the disposition of SRSM will have different environmental impacts. For example, if conditional clearance is chosen, the use of landfill disposal at sites near nuclear power plants will reduce transportation and associated vehicle emissions. These impacts may be small relative to the potential radiation-related impacts on health and the economic impacts, but they must be examined to ensure that any regulation does not produce worse environmental impacts as an unintended consequence.

Direct Costs

The main direct cost impact of alternative approaches is likely to be the licensees' disposal costs for SRSM. A no-release policy means, in practice, that all low-level radioactive materials would have to be sent to a site licensed to accept LLRW for disposal. If conditional clearance is chosen, the cost of disposal of metals at a landfill site, even a Subtitle C hazardous waste landfill, is substantially lower than the cost at LLRW sites. The committee's preliminary calculations (Chapter 4) indicate that disposal of decommissioning wastes under a strict no-release policy would cost *billions* of dollars, whereas Subtitle D landfill disposal would cost a few hundred *million* dollars.

The committee reviewed available cost data but found only limited information. The current cost estimates of disposal vary widely, both among LLRW sites and between LLRW sites and landfill disposal options. Because cost will be a major factor in selecting an approach for disposing of SRSM, it is very important that the USNRC conduct a thorough cost analysis that accounts for the differences among disposal options and the uncertainties in cost estimates caused by regulations and by supply and demand.

Other waste management costs will include transportation and operational (e.g., material preparation and sample analysis) costs. These are likely to be much lower than disposal costs. Regulatory costs also have to be considered. These include the cost of staff at the USNRC and in agreement states to manage whichever regulatory approach is taken.

Indirect Costs

Indirect costs of alternative approaches include the potential liabilities of licensees and other waste handlers. Although the study committee has not heard of any cases where such liabilities were invoked, some industry representatives clearly expressed concerns about this possibility. However an approach is fashioned, it must consider the liability of generators in a variety of circumstances, including continuing liability, erroneous free release, and unapproved reuse.

As noted above, representatives from the steel and concrete industries have expressed particular concern about the impact of releasing slightly radioactive steel and concrete into commerce. They believe that the presence of released material in their feed streams could stigmatize their products, reducing sales and revenue. Representatives of these industries made it quite clear that their policy is to reject any materials identified as radioactive by detection equipment at their gates when the material arrives at their facilities. They emphasized that their companies will continue to exercise vigilance in this area.

Direct Benefits

Alternatives allowing clearance would create opportunities for commercial benefits—for example, through the sale of SRSM. One example is the sale being contemplated by the DOE of $30 million worth of slightly radioactive nickel on the commercial market. The committee did not hear much evidence for potential direct benefits (other than the nickel example), but it would be useful to determine the net value associated with releasing these materials into commerce. These net value calculations should consider both the market value of the materials and the cost of processing and shipping them. Another direct benefit is the reduction of licensees' operational expenses. For example, licensees expressed concern about the paperwork and cost of releasing equipment to be moved from one controlled site to another, but they did not comment on additional potential labor costs associated with further categorization of waste materials.

Consistency with Existing Regulations

Consistency with international, national, state, and local regulations is desirable, even though it should not be the main reason for selecting an alternative. In Chapter 7, the committee discusses the efforts under way in the European Union (EU) to establish consistent standards for free release of SRSM. There may be economic advantage to the United States in establishing a clearance standard for SRSM, particularly if it were consistent with international standards. Consistency would make import-export and control of materials easier and, if monitored properly, of no consequence to public health. An international agreement on such trade not only must include the levels of residual radioactivity allowable for clearance for shipment, but also must specify standard methodologies of measurement at both the point of export and the point of import. Standard measurement methods are particularly important for ensuring detection, and preventing the shipment, of materials in which orphan sources are present.

The committee believes that the USNRC may wish to evaluate the various technical considerations employed by the EU and other countries in reaching clearance standards. However, as stated elsewhere in this report, the committee believes that many other factors should be considered in any U.S. approach.

Consistency with other federal regulations is also important. For example, the rulemaking process employed by the Environmental Protection Agency (EPA) results in lengthy explanation of all comments in the preamble to the *Federal Register* announcement of the rule. Under RCRA, the EPA establishes acceptable risk levels and then develops compound-by-compound standards through detailed calculations for each chemical and environmental medium. The EPA approach results in a detailed explanation of regulatory decisions, aspires to consistent application of risk, and elicits extensive public participation. It also includes extensive responses and analyses of public comments in *Federal Register* announcements as well as in administrative records.

Similarly, if the USNRC were to choose a dose-based approach to setting a national standard, consistency with the regulation of other radioactive materials would be important. For example, the committee is concerned about inconsistencies with the current regulatory approaches to NORM, technologically enhanced naturally occurring radioactive material (TENORM), and NARM wastes. The issue of consistency within USNRC guidelines and regulations should be addressed as well.

Implementation and Enforcement

To be effective, any approach to clearance of SRSM must be implementable and enforceable. Of special relevance in this case is the ability to detect, measure, and monitor very small amounts of radiation with few false alarms. Another concern is the ability to track the chain of custody of conditionally cleared materials, especially if the uses of these materials are restricted by conditions on their release. Hence, to establish confidence in any approach to clearance of SRSM, there must be adequate procedural guidance, oversight, and reporting requirements.

Enforceability is crucial for ensuring broad-based compliance with a standard. Both enforceability and a standardized, accessible measurement methodology are crucial for uniform implementation. Enforceability requires penalties (such as fines) for failure to meet the standard or failure to implement the standard properly. Enforcement by regulatory agencies is an integral part of gaining public trust as well.

Public Perception

The USNRC faces perhaps no greater challenge than winning widespread public acceptance of any regulation for release of SRSM. As discussed in the next section of this chapter, there are many challenges, opportunities, and options for the USNRC in seeking public acceptance. Acceptance does not equate directly with consensus or unanimous agreement. Rather, the likelihood of acceptance is increased first by adhering faithfully to an announced process that engages all responsible stakeholder representatives and viewpoints. Second, this process must be perceived by participants as fair and open. Third, the process should bring out all advantages and disadvantages of the alternative approaches in an even-handed way. Fourth, participation throughout the process by informed and knowledgeable persons, as well as openness to a broad and creative range of alternatives, will increase public acceptance.

The USNRC could use many mechanisms to attain public acceptance. The committee believes that the degree of trust (or mistrust) of the USNRC has been and will remain a major factor in the public's response to issues involving SRSM. The USNRC should consider substantial changes that would open its decision-making process (for details, see "Stakeholder Involvement" in Chapter 2 and all of Chapter 8).

Decision Impact Matrix

Figure 9-1 shows, in the form of a two-dimensional matrix, the committee's view of how alternative approaches and their possible impacts should be analyzed and evaluated. A thorough and systematic analysis and evaluation of these approaches would address each cell of this matrix. Additional columns and rows might emerge from a thorough stakeholder involvement process.

Most of the work to date on evaluating alternatives has focused on health impacts. Although this is an important issue when setting a standard, other impacts may be significant as well. The committee has done some preliminary work on some of these other impacts. For example, the relative costs for a conditional clearance standard and a no-release alternative are discussed in Chapter 4. However, there clearly is much more work to be done to provide a satisfactory assessment for all of the alternatives and impacts represented in Figure 9-1.

FINDINGS

Finding 9.1. The committee found no evidence that the problems with the current approach to clearance decisions require its immediate replacement. The committee concludes that there is sufficient time to conduct a thorough and systematic analysis and evaluation, including a sound process of stakeholder participation and involvement, of alternative approaches to the disposal of SRSM.

Finding 9.2. Although there are many possible alternatives for the disposal of SRSM from USNRC-licensed facilities, the committee heard substantial support

	Case-by-Case Approach	Clearance Standard	Conditional Clearance Standard			No Releases
	RG 1.86 or modification	No restrictions on reuse	Beneficial reuse in controlled environments	Reuse for low-exposure scenarios	Landfill disposal	LLRW disposal
Health Impacts						
Environmental Impacts						
Direct Costs						
Indirect Costs						
Direct Benefits						
Consistency with Existing Regulations						
Implementation and Enforcement						
Public Perception						

FIGURE 9-1 Decision impact matrix. This matrix represents impacts to be assessed in evaluating alternatives for the disposition of SRSM from USNRC-licensed facilities. The alternatives shown as the column headings are preliminary selections by the study committee for this report; a different set might emerge from a thorough approach to public involvement at all levels of the decision process. NOTE: RG = Regulatory Guide.

from stakeholders for only a few. In general terms, the supported alternatives are a dose-based clearance standard, a dose-based conditional clearance standard, and a no-release policy. Different stakeholders expressed preferences for different conditions for a dose-based conditional clearance standard: beneficial reuse in controlled environments, commercial reuse in low-exposure scenarios, or landfill disposal. Source-based standards and minor modifications of the existing case-by-case approach received limited support.

Finding 9.3. There are many possible impacts of the approaches that the USNRC might select for the clearance of SRSM. Potentially important impacts include the degree of public protection against exposure from radioactive materials, environmental impacts, direct costs (e.g., for disposal), indirect costs (e.g., through product stigmatization), consistency with existing regulations, implementation and enforcement, and public perception. To date, the USNRC has focused its analyses of alternative approaches fairly narrowly on protecting the public from exposure to SRSM. The USNRC has done very little analysis of the other important impacts on this list.

10

Findings and Recommendations

The U.S. Nuclear Regulatory Commission's (USNRC's) regulations on protection against radiation, 10 CFR Part 20, do not contain predetermined concentrations, amounts, or quantities of radionuclides in solid materials below which these materials can be released from further regulatory control. Solid materials potentially available for release from regulatory control include metals, building concrete, on-site soils, equipment, and furniture used in routine operation of licensed nuclear facilities. Most of this material will have no radioactive contamination, but some of it may have surface or volume contamination. Licensees continue to request permission from the USNRC and agreement states to release such solid materials when the materials are no longer useful, pursuant to Section 2002 of 10 CFR Part 20 or compatible state regulations, or when the licensed facility is decommissioned. The USNRC does use a guidance document issued by the Atomic Energy Commission in 1974, Regulatory Guide 1.86, which contains limits applicable to surface contamination and allows clearance of solid materials, usually by incorporation into license technical specifications.

The USNRC allows licensees to release solid material according to preestablished criteria. For reactors, if surveys for surface residual radioactivity performed by the licensee on equipment or materials indicate the presence of radioactivity above natural background levels then release is not permissible.[1] If no such surface activity is detected, then the solid material in question need not be treated as waste under 10 CFR Part 20. This approach sometimes leads to prob-

---

[1] Reactor licensees can apply to USNRC for approval for clearance of solid materials with small but detectable levels of radioactivity pursuant to Section 2002 of 10 CFR Part 20 on a case-by-case basis.

lems when detectors of greater sensitivity than were used in the initial survey detect radioactivity above the threshold in previously released material (USNRC, 2001b). For surface-contaminated solid materials possessed by a materials licensee, the USNRC usually authorizes the release through specific license conditions (USNRC, 2001b). In the case of volume-contaminated materials held by reactor and materials licensees, the USNRC has not provided guidance similar to that found in Regulatory Guide 1.86 for surface contamination. These situations are instead decided on an individual basis by evaluating the doses likely to be associated with the proposed disposition of the material.

The USNRC has attempted to update and formalize its policies on disposition of slightly radioactive solid material (SRSM). In 1990, the USNRC issued a policy as directed by the Low Level Radioactive Waste Policy Amendments Act of 1985 (LLWPAA) that declared materials with low concentrations of radioactivity contamination to be "below regulatory concern" (BRC) and hence deregulated (55 Federal Register 27522; July 3, 1990). However, Congress intervened to set aside the BRC policy in the Energy Policy Act of 1992 after the USNRC's own suspension of the policy (56 Federal Register 36068; July 30, 1991). In 1999, the USNRC again examined the issue of disposition of SRSM and published a *Federal Register* notice examining several policy options (64 Federal Register 35090-35100; June 30, 1999). In neither case was the USNRC able to convince consumer and environmental groups that clearance of SRSM could be done safely or to convince some industry groups that clearance is desirable.

In August 2000, the USNRC asked the National Research Council to form a committee to provide advice in a written report. The committee addresses its tasks in Chapters 2 through 9 of the report, each of which contains a set of findings, a subset of which is presented. The reader is encouraged to review all of the findings as well as the supporting documentation in each chapter. The major findings and recommendations follow.

MAJOR FINDINGS

Regulatory Framework (Chapter 2)

Finding 2.1. The USNRC does not have a clear, overarching policy statement for management and disposition of SRSM. However, SRSM has been released from licensed facilities into general commerce or landfill disposal for many years pursuant to existing guidelines (e.g., Regulatory Guide 1.86) and/or following case-by-case reviews. The USNRC advised the committee of no database for these releases.

Finding 2.2. A dose-based clearance standard can be linked to the estimated risk to an individual in a critical group from the release of SRSM. The general regulatory trend is toward standards that are explicitly grounded in estimating risks.

Finding 2.3. For clearance of surface-contaminated solid materials, the clearance practices regulated by the USNRC and agreement states are based on the guidance document Regulatory Guide 1.86, which is technology based and has been used satisfactorily in the absence of a complete standard since 1974.

Finding 2.4. For clearance of volume-contaminated solid materials, the USNRC has no specific standards in guidance or regulations. Volume-contaminated SRSM is evaluated for clearance on a case-by-case basis. This case-by-case approach is flexible, but it is limited by outdated, incomplete guidance, which may lead to determinations that are inconsistent.

Finding 2.5. Industrial activities are generating very large quantities of technologically enhanced naturally occurring materials (TENORM). Federal regulation of TENORM has been largely absent. State regulations vary in breadth and depth.

Anticipated Inventories of Radioactive or Contaminated Materials (Chapter 3)

Finding 3.1. Licensees may seek to clear about 740,000 metric tons of metallic SRSM that arise from decommissioning the current population of U.S. power reactors during the period 2006 to 2030 (about 30,000 to 42,000 metric tons per year). About 8,500 metric tons per year are expected to arise from decommissioning USNRC-licensed facilities other than power reactors during the same time period. The total quantity of metal from both power reactor and non-power reactor licensees, up to approximately 50,000 metric tons per year, represents about 0.1 percent of the total obsolete steel scrap that might be recycled during that same 25-year period.

Finding 3.2. If most of the licensees of currently operating reactors obtain 20-year license extensions, relatively little SRSM will arise from power plant decommissioning during the 2006-2030 period.

Finding 3.3. Because of the difficulty of determining the quantities and levels of contamination that have penetrated into the concrete, concrete SRSM is generally considered to be volume contaminated. Concrete SRSM constitutes more than 90 percent of the total SRSM arising from decommissioning the population of U.S. power reactors.

Pathways and Estimated Costs for Disposition of Slightly Radioactive Solid Materials (Chapter 4)

Finding 4.1. Disposal of all slightly radioactive solid materials arising from decommissioning the population of U.S. power reactors into low-level radioactive waste disposal sites would be expensive (about $4.5 billion to $11.7 billion) at current disposal charge rates. Disposal in Subtitle D or Subtitle C landfills would be cheaper ($0.3 billion to $1 billion, respectively). Clearance of all of this material could reduce disposal costs to nearly zero (assumes 100 percent reuse or recycle) or might even result in some income (~$20 million) arising from the sale of scrap materials for recycle or reuse. Decontamination, segmentation, and transport costs are not included in the costs estimated in this report for disposition.

Review of Methodology for Dose Analysis (Chapter 5)

Finding 5.1. Analytical work in the United States and abroad over the past two decades is useful in understanding the likely doses associated with exposure scenarios that might occur under various clearance standards. Much of the technical analysis in this field has the objective of understanding "dose factors," which to date have been analyzed in depth only for (unconditional) clearance scenarios. A dose factor is used to convert a concentration of radioactivity that is about to be released, whether it be confined to a surface or contained within a volume, to a primary dose level (measured in microsieverts per year or millirems per year). With such a dose factor in hand, a primary dose standard can be converted to obtain a secondary clearance standard in terms of radionuclide activity, which could then be used at USNRC-licensed facilities. A dose factor can be used with any choice of primary dose standard.

Finding 5.2. Selecting a primary dose standard is a policy choice, albeit one informed by scientific estimates of the health risk associated with various doses. For instance, as shown in Table 1-2, a lifetime dose rate of 10 μSv/yr (1 mrem/yr) equates to an estimated increased lifetime cancer risk of 5×10^{-5}, which falls within the range of acceptable lifetime risks of 5×10^{-4} to 10^{-6} used in developing health-based radiation standards other than radon in the United States (NRC, 1995, p. 50). When setting primary dose standards, regulators can make a policy decision to include a level of conservatism such that the final standard is in excess of the best-estimate dose factor and in this way account for uncertainty (e.g., selecting the 90th, 95th, or other percentile in the distribution for the dose factor, instead of the best-estimate value).

Finding 5.3. The uncertainty in dose factor estimates is a key technical issue. When an uncertainty has been estimated, a quantitative determination can be made of the likelihood that the dose to an individual in the critical group will be below the primary dose standard. Quantitative uncertainty estimates can also assist regulators in assigning a level of conservatism to dose factors in excess of the best estimate. Dose factors developed by analysts from different countries show wide variation, which highlights the need for careful consideration of uncertainties.

Finding 5.4. The committee concludes from its review that of the various reports, draft NUREG-1640 (USNRC, 1998b) provides a *conceptual framework* that best represents the current state of the art in risk assessment, particularly with regard to its incorporation of formal uncertainty, as judged using recommendations of this committee and other committees of the National Research Council. Once the limitations in draft NUREG-1640 have been resolved (see Findings 5.5 and 5.6 [see Chapter 5]) and the results are used in conjunction with appropriate dose-risk estimates—in the final version of the report or in follow-up reports—the USNRC will have a sound basis for considering the risks associated with any proposed clearance standards and for assessing the uncertainty attached to these dose estimates.

Finding 5.7. The dose factors developed in draft NUREG-1640 should not be used to derive clearance standards for categories of SRSM other than those considered in the draft NUREG-1640, without first assessing the appropriateness of the underlying scenarios. Some of the dose factors developed in draft NUREG-1640 are likely to require modification when applied to other mixtures of radionuclides (e.g., mixtures in which transuranics dominate) and other clearance scenarios, such as may be relevant to DOE material and technologically enhanced naturally occurring radioactive material (TENORM).

Measurement Issues (Chapter 6)

Finding 6.3. For a 1 mrem/yr or higher standard (and the corresponding derived secondary screening levels), the majority of radionuclides can be detected at reasonable costs in a laboratory setting, under most practical conditions. For a 0.1 mrem/yr standard, the measurement capability falls below the upper bound of minimum detectable concentrations for some radionuclides in some laboratories, although 85 percent of radionuclides are still detectable. Using field measurements, a more rapid fall-off of detectability is observed at more stringent radiation protection levels, with 31 of 40 key radionuclides detectable at 1 mrem/yr and 11 of 40 detectable at 0.1 mrem/yr.

International Approaches to Clearance (Chapter 7)

Finding 7.1. The EU and the IAEA have each established a dose-based standard of 10 µSv/yr (1 mrem/yr) for the clearance of materials from regulatory control. A collective dose standard is also included, expressed as a committed dose equivalent of 1 man-Sv per year of exposure of the affected group (100 man-rem total effective dose equivalent per year).[2]

---

[2]Provisions exist in the EU safety directives for competent authorities in member states to develop alternative clearance guidance for special or specific circumstances.

FINDINGS AND RECOMMENDATIONS   171

Finding 7.3. A body of science, policy, and literature supports the development of the EU safety directives related to radioactive solid material clearance. In particular, the IAEA has developed policy guidance found in *Principles for the Exemption of Radiation Sources and Practices from Regulatory Control* (IAEA, 1988).

Stakeholder Reactions and Involvement (Chapter 8)

Finding 8.1. The USNRC involved stakeholders in the processes for the BRC policy and the License Termination Rule for decommissioning, as well as in the initial stages of considering standards for release of SRSM. Despite these efforts, environmental and consumer advocacy groups remain concerned with radiation effects, and industrial groups continue to be concerned with the potential economic consequences of the clearance of SRSM.

Finding 8.3. Stakeholder groups differed in their viewpoints on regulating disposition of SRSM. Generally, professional societies associated with the nuclear industry supported clearance, industrial groups endorsed conditional clearance, and environmental groups opposed any type of clearance. However, much of the opposition to a clearance standard was associated with recycling metal SRSM into general commerce.

Finding 8.4. A legacy of distrust of the USNRC has developed among most of the environmental stakeholder groups. This distrust results from their experience with the BRC policy, the License Termination Rule, and the 1999 issues paper on the release of SRSM. Reestablishing trust will require concerted and sustained effort by the USNRC, premised on a belief that stakeholder involvement will be important and worthwhile, as well as a prerequisite for making progress.

Framework and Process for Decision Making (Chapter 9)

Finding 9.1. The committee found no evidence that the problems with the current approach to clearance decisions require its immediate replacement. The committee concludes that there is sufficient time to conduct a thorough and systematic analysis and evaluation, including a sound process of stakeholder participation and involvement, of alternative approaches to the disposal of SRSM.

RECOMMENDATIONS

In developing its recommendations the committee was guided by two overarching, compelling findings:

1. The current approach to clearance decisions is workable and is sufficiently protective of public health that it does not need immediate re-

vamping. However, the current approach, among other shortcomings, is inconsistently applied, is not explicitly risk based, and has no specific standards in guidance or regulations for clearance of volume-contaminated slightly radioactive solid material. Therefore, the committee believes that the USNRC should move ahead without delay and start a process of evaluating alternatives to the current system and its shortcomings.

2. Broad stakeholder involvement and participation in the USNRC's decision-making process on the range of alternative approaches is critical as the USNRC moves forward. The likelihood of acceptance of a USNRC decision greatly increases when the process (1) engages all responsible stakeholder representatives and viewpoints, (2) is perceived by participants as fair and open, (3) addresses all the advantages and disadvantages of the alternative approaches in an even-handed way, and (4) is open to a broad and creative range of alternatives. Thus, it is essential that the USNRC focus on the process and not prescribe an outcome. The outcome, an approach to disposition of slightly radioactive solid material, must evolve from the process.

While the committee did not want to prescribe the outcome of the decision process, it has made several specific recommendations, conditional on the process arriving at certain decision points. For example, if the USNRC contemplates clearance or conditional clearance standards, the committee recommends that these standards be dose based. The committee also recognized that significant national and international efforts have been completed, or are near completion, that provide a solid foundation for the USNRC to move forward. The committee has recommended the foundation from which to begin the process. Thus, the USNRC should be able to proceed expeditiously with a broad-based stakeholder participatory decision making process.

Recommendation 1. The USNRC should devise a new decision framework that would develop, analyze, and evaluate a broader range of alternative approaches to the disposition of slightly radioactive solid material. At a minimum, these alternatives should include the current case-by-case approach, clearance, conditional clearance, and no release.

Recommendation 2. The USNRC's decision-making process on the range of alternative approaches to the disposition of slightly radioactive solid material should be integrated with a broad-based stakeholder participatory decision-making process. Elements of this process should include the following:

- The willingness and commitment of the USNRC to establish and maintain a meaningful and open dialogue with a wide range of stakeholders regarding the disposition of slightly radioactive solid material;

• An ad hoc broad-based advisory board that would advise the USNRC in its consideration of approaches to the disposition of slightly radioactive solid material. The advisory board would also suggest additional stakeholder involvement mechanisms that the USNRC could use in the decision process (for example, establishing a National Environmental Policy Act process; alternative dispute resolution; and partnering, arbitration, mediation, or a combination of such methods); and

• Assistance obtained by the USNRC as needed from outside experts in order to (1) assist its efforts to establish the ad hoc stakeholder advisory board and to facilitate dialogue among the USNRC and stakeholder participants in the decision-making process and (2) assess, evaluate, and perhaps conduct portions of the USNRC stakeholder involvement program and make recommendations as appropriate.

Recommendation 3. The USNRC should adopt an overarching policy statement describing the principles governing the management and disposition of slightly radioactive solid material. A review and discussion of the IAEA policy statement *Principles for the Exemption of Radiation Sources and Practices from Regulatory Control* (Safety Series No. 89, IAEA Safety Guidelines, Vienna, 1988) with a broad-based stakeholder group would provide a good starting point in developing a policy statement that would provide a foundation for evaluation of alternative approaches to disposition of slightly radioactive solid material.

Recommendation 4. When considering either clearance or conditional clearance, a dose-based standard should be employed as the primary standard. To employ a dose-based standard, it is necessary to consider a wide range of scenarios that encompass the people likely to be exposed to slightly radioactive solid material. From these people, a critical group is selected and secondary standards (based on dose factors) are derived. These secondary standards are used to limit the radioactivity in materials being considered for release or conditional release.

The USNRC should also consider the pros and cons of the establishment of a separate collective dose standard.

Recommendation 5. An individual dose standard of 10 µSv/yr (1 mrem/yr) provides a reasonable starting point for the process of considering options for a dose-based standard for clearance or conditional clearance of slightly radioactive solid material. This starting point is appropriate for the following reasons:

• A dose of 10 µSv/yr (1 mrem/yr) is a small fraction (less than 0.5 percent) of the radiation received per year from natural background sources.

- A dose of 10 µSv/yr (1 mrem/yr) is significantly less than the amount of radiation that we receive from our own body due to radioactive potassium (one contributor to background radiation) and other elements and to routine medical procedures that involve ionizing radiation.
- A dose of 10 µSv/yr (1 mrem/yr) over a 70-year lifetime equates to an estimated increase of 3.5×10^{-5} in the lifetime cancer risk, which falls within the range of acceptable lifetime risks of 5×10^{-4} to 10^{-6} used in developing health-based standards for exposure to radiation (other than for radon) in the United States.
- Radiation measurement technologies are available at a reasonable cost to detect radioactivity at concentrations derived from this dose standard.
- This dose standard is widely accepted by recognized national and international organizations.

The final selection of an individual dose standard should nonetheless be a policy choice, albeit one informed by the above considerations.

Recommendation 6. For any dose-based alternative approach to disposition of slightly radioactive solid materials, the USNRC should use the *conceptual framework* of draft NUREG-1640 to assess dose implications. To use the actual results of NUREG-1640 in the decision framework discussed in Recommendations 1 and 2, the USNRC must first establish confidence in the numerical values, expand the scope of applicability, and overcome certain limitations in draft NUREG-1640. At a minimum, the following specific actions are required:

- Review the choice of parameter distributions used in the dose modeling, as well as the characteristic values chosen for each parameter distribution.
- Develop complete scenarios and dose factors for conditional clearance options.
- Provide sufficient information to enable calculation of collective doses to support Recommendation 4.
- Expand the current set of scenarios used to compute dose factors to include (1) human error and (2) multiple exposure pathways.

The USNRC should use an independent group of experts to provide peer review of these activities.

Recommendation 7. The USNRC should continue to review, assess, and participate in the ongoing international effort to manage the disposition of slightly radioactive solid material. The USNRC should also develop a rationale for consistency between secondary dose standards that may be adopted by the United States and other countries. However, the USNRC should ensure that the technical basis for secondary dose standards is not adjusted for consistency unless these adjustments are supported by scientific evidence.

References

American National Standards Institute and Health Physics Society (ANSI/HPS). 1999. Surface and Volume Radioactivity Standards for Clearance. ANSI/HPS N13.12-1999. Washington, D.C.: American National Standards Institute, Inc.

Atomic Energy Commission (AEC). 1974. Termination of Operating Licenses for Nuclear Reactors. Regulatory Guide 1.86. Washington, D.C.

Bier, V. 1999a. Summary of the State of the Art: Risk Communication to the Public. August. University of Wisconsin-Madison.

Bier, V. 1999b. Summary of the State of the Art: Risk Communication to Decision Makers. December. University of Wisconsin-Madison.

Bolch, W. E., et al. 2001. Influences of Parameter Uncertainties Within the ICRP 66 Respiratory Tract Model: Particle Deposition. Health Physics 81(4):378-394.

Breshears, D. D., et al. 1989. Uncertainty in Predictions of Fallout Radionuclides in Foods and of Subsequent Ingestion. Health Physics 57(6):943-953.

Center for Nuclear Waste Regulatory Analysis (CNWRA). 2001. Review of Draft NUREG-1640, Radiological Assessment of Clearance of Equipment and Materials from Nuclear Facilities. April 21, San Antonio, Tex.: CNWRA.

Clarke, R.H. 2001. Exclusion, Exemption and Clearance . . . Do We Need Them All? Health Physics 81(2):105.

Conference of Radiation Control Program Directors (CRCPD). 1997. Part N—Regulation and Licensing of Technologically Enhanced Naturally Occurring Radioactive Materials (TENORM). Draft. Frankfort, Ky: CRCPD.

Cooke, R. 1991. Experts in Uncertainty: Opinion and Subjective Opinion in Science. New York: Oxford University Press.

Cox, F.M., and C.S. Guenther. 1995. An Industry Survey of Current Lower Limits of Detection for Various Radionuclides. Health Physics 69: 21-129.

Creighton, J., and J. Priscoli. 1996. Overview of Alternative Dispute Resolution (ADR): A Handbook for Corps Managers. 96-ADR-P-5. Prepared for the U.S. Army Corps of Engineers. Washington, D.C.: U.S. Army Corps of Engineers.

Department of Energy (DOE). 1993a. Radiation Protection of the Public and the Environment. DOE 5400.5. Office of Environment, Safety and Health. Washington, D.C.: DOE.

DOE. 1993b. Earning Public Trust and Confidence: Requisites for Managing Radioactive Waste. Advisory Board Task Force on Radioactive Waste Management. Washington, D.C.: DOE.

DOE. 1996. The 1996 Baseline Environmental Management Report. Volume 1. DOE/EM-0290. June. Washington, D.C.: DOE.

DOE. 1999. Reuse of Concrete from Contaminated Structures, DOE/OR/22343-1. January. Washington, D.C.: DOE.

DOE. 2000. Audit Report: The Decontamination and Decommissioning Contract at the East Tennessee Technology Park. DOE/IG-0481. Washington, D.C.: DOE.

DOE. 2001. Notice of intent to prepare a programmatic environmental impact statement on the disposition of scrap metals and announcement of public scoping meetings. Federal Register 67:36562–36566, July 12.

Elder, H.K. 1981. Technology, Safety and Costs of Decommissioning a Reference Uranium Hexafluoride Conversion Plant. NUREG/CR-1757, Pacific Northwest Laboratory for U.S. Nuclear Regulatory Commission. Washington, D.C.: U.S. Nuclear Regulatory Commission.

Envirocare. 2001. Envirocare of Utah, Inc., Waste Acceptance Guidelines, May 16, Revision 3. Clive, Utah: Envirocare.

Environmental Protection Agency (EPA). 1989. Environmental Standards for Management, Storage, and Land Disposal of Naturally Occurring and Accelerator-Produced Radioactive Waste. Draft Rule for 40 CFR. Part 764, never finalized. April 6. Washington, D.C.: EPA.

EPA. 1996. Summary Report for the Workshop on Monte Carlo Analysis. Risk Assessment Forum. Washington, D.C.: EPA.

EPA. 1997a. Technical Support Document: Evaluation of the Potential for Recycling of Scrap Metals from Nuclear Facilities. TSD 97. Peer Review Draft, March 11. Office of Radiation and Indoor Air. Washington, D.C.: EPA.

EPA. 1997b. Guidance: Establishment of Cleanup Levels for CERCLA Sites with Radioactive Contamination. OSWER No. 9200.4-18. Office of Solid Waste and Emergency Response. Washington, D.C.: EPA.

EPA. 1997c. Radiation Protection Standards for Scrap Metal: Preliminary Cost-Benefit Analysis. Clean Materials Program, Report No. Washington, D.C.: U.S. Environmental Protection Agency.

EPA. 1997d. Letter from Carol M. Browner, administrator, U.S. Environmental Protection Agency, to the Honorable Shirley Jackson, chair, U.S. Nuclear Regulatory Commission, February 7.

EPA. 2000. Draft Public Involvement Policy. Federal Register 65: 82335–82345, December 28.

EPA. 2001. Radiologically Contaminated Superfund National Priorities List Sites. Available at <www.epa.gov/radiation>. Accessed March 1, 2002.

EPA, USNRC, and DOE. 2000. Multi-agency Radiation Survey and Site Implementation Manual. NUREG-1557, Revision 1; EPA 402-R-97-016, Revision 1; DOE/EH-0624, Revision 1 (August). Washington, D.C.: EPA, USNRC, and DOE.

European Commission (EC). 1998a. Communication from the Commission Concerning the Implementation of Council Directive 96/29/Euratom Laying Down Basic Safety Standards for the Protection of the Health and Safety of Workers and the General Public Against the Dangers Arising from Ionising Radiation. Luxembourg: European Communities.

EC. 1998b. Radiation Protection 89: Recommended radiological protection criteria for the recycling of metals from dismantling of nuclear installations. Recommendations of the group of experts set up under the terms of Article 31 of the Euratom Treaty. Luxembourg: European Communities.

EC. 2000. Definition of Clearance Levels for the Release of Radioactively Contaminated Buildings and Building Rubble. EC-RP-114. Luxembourg: European Communities.

REFERENCES

EC. 2001. Radiation Protection 122: Practical use of the concepts of clearance and exemption. Part 1. Guidance on general clearance levels for practices. Luxembourg: European Communities.

European Union (EU). 1992. Council Directive 92/3/Euratom on the Supervision and Control of Shipments of Radioactive Waste between Member States and into and out of the Community (OJ L 35, 12.2.92), February 12. Luxembourg: European Communities.

EU. 1996. Council Directive 96/29/EURATOM (OJ L 159, 29.6.1996), May 13. Luxembourg: European Communities.

Flynn, J., W. Burns, C. Mertz, and P. Slovic. 1992. Trust as a Determinant of Opposition to a High-Level Radioactive Waste Repository: Analysis of a Structural Model. Risk Analysis 12:417-430.

Haimes, Y.Y. 1991. Total Risk Management. Risk Analysis 11:169-171.

Inside NRC. 2001. DOE Ends Contract with SAIC for EIS on Releases of Solid Materials, July 30. New York: Platts.

International Atomic Energy Agency (IAEA). 1988. Principles for the Exemption of Radiation Sources and Practices from Regulatory Control. Safety Series No. 89. Vienna: IAEA.

IAEA. 1992. Application of Exemption Principles to the Recycle and Reuse of Materials from Nuclear Facilities. Safety Practice No. 111-P-1.1. Vienna: International Atomic Energy Agency.

IAEA. 1996. Clearance Levels for Radionuclides in Solid Materials: Applications of Exemption Principles. Interim Report. Vienna: International Atomic Energy Agency.

International Commission on Radiological Protection (ICRP). 1985. Statement from 1985 Paris Meeting of the ICRP. British Journal of Radiology (58) 910.

ICRP. 1990. Recommendations of the International Commission on Radiological Protection. Issue No. 60, Vol. 21/1-3. Oxford, England: Pergamon Press.

Jefferson, T. 1820. Letter to William Charles Jarvis, September 28.

Johnston, G. 1991. An Evaluation of Radiation and Dust Hazards at a Mineral Sand Processing Plant. Health Physics 60(6):781-787.

Kasperson, R., O. Renn, P. Slovic, H. Brown, J. Emel, R. Bogel, T.J. Kasperson, and S. Patrick. 1988. The Social Amplification of Risk: A Conceptual Framework. Risk Analysis 8:177-187.

Konzek, G., et al. 1995. Revised Analyses of Decommissioning for the Reference Pressurized Water Reactor Power Station, NUREG/CR-5884 (November). Washington, D.C.: U.S. Nuclear Regulatory Commission.

Lubenau, J. 1998. Radioactive Materials in Recycled Metals–An Update. Health Physics 74(3):293-299.

Mariotte, M. et al. 1995. Letter to Hon. Shirley Jackson, chair, U.S. Nuclear Regulatory Commission, September 25, 1995.

Morgan, M., and M. Henrion. 1990. Uncertainty: A Guide to Dealing with Uncertainty in Quantitative Risk and Policy Analysis. Cambridge, England: Cambridge University Press.

National Council on Radiation Protection and Measurements (NCRP). 1987a. Ionizing Radiation Exposure of the Population in the United States. Report No. 93. Bethesda, Md.: NCRP.

NCRP. 1987b. Exposure of the Population in the United States and Canada from Natural Background Radiation. Report No. 94. Bethesda, Md.: NCRP.

NCRP. 1987c. Recommendations on Limits for Exposure to Ionizing Radiation. Report No. 91. Bethesda, Md.: NCRP.

NCRP. 1987d. Radiation Exposure of the United States Population from Consumer Products and Miscellaneous Sources, Report No. 95. Bethesda, Md: NCRP.

NCRP. 1989a. Exposure of the United States Population from Diagnostic Medical Radiation, Report No. 100, 1999. Bethesda, Md: NCRP.

NCRP. 1989b. Radiation Protection for Medical and Allied Health Personnel, NCRP Report No. 105, 1989. Bethesda, Md: NCRP.

NCRP. 1993. Limitation of Exposure to Ionizing Radiation. Report No. 116. Bethesda, Md.: NCRP.

NCRP. 1998. Operational Radiation Safety Program. Report No. 127. Bethesda, Md.: NCRP.
National Research Council (NRC). 1989. Improving Risk Communication. Committee on Risk Perception and Communication. Washington, D.C.: National Academy Press.
NRC. 1990. Health Effects of Exposure to Low Levels of Ionizing Radiation. BEIR V. Committee on the Biological Effects of Ionizing Radiation. Washington, D.C.: National Academy Press.
NRC. 1994. Science and Judgement in Risk Assessment. Washington, D.C.: National Academy Press.
NRC. 1995. Technical Basis for Yucca Mountain Standards. Washington, D.C.: National Academy Press.
NRC. 1996. Affordable Cleanup? Opportunities for Cost Reduction in the Decontamination and Decommissioning of the Nation's Uranium Enrichment Facilities. Committee on Decontamination and Decommissioning of Uranium Enrichment Facilities, Board on Energy and Environmental Systems. Washington, D.C.: National Academy Press.
NRC. 1999. Evaluation of Guidelines for Exposures to Technologically Enhanced Naturally Occurring Radioactive Materials. Committee on Evaluation of EPA Guidelines for Exposure to Naturally Occurring Radioactive Materials, Board on Radioactive Effects Research, Commission on Life Sciences, National Research Council. Washington, D.C.: National Academy Press.
NRC. 2001. The Impact of Low-Level Radioactive Waste Management Policy on Biomedical Research in the United States. Washington, D.C. National Academy Press.
Natural Resources Defense Council (NRDC). 1991. Letter of November 12, 1991, from D. Reicher and T. Cochran, NRDC, to U.S. Nuclear Regulatory Commission commissioner Rogers.
Newberry, B. 2001. Letter to Barnwell Customers from Bill Newberry, manager, Radioactive Waste Disposal Program, State Budget and Control Board, South Carolina, August 1.
Omenn, G. 1997. Framework for Environmental Health Risk Management. Commission on Risk Assessment and Risk Management. Final Report. Volume 1. Washington D.C.: U.S. Government Printing Office.
Pescatore, C. 2001. Management of Slightly Contaminated Materials: Status and Issues. Paper presented at VALDOR Conference, Stockholm, June.
Pijawka, K., and A. Mushkatel. 1992. Public Opposition to the Siting of the High-level Nuclear Waste Repository: The Importance of Trust. Policy Studies Review 10(4): 180-194.
Raynor, S., and R. Cantor. 1987. How Fair Is Safe Enough: The Cultural Approach to Societal Technological Choice. Risk Analysis 7:3-9.
Sanford Cohen & Associates, Inc. (SCA). 2001. Inventory of Materials with Very Low Levels of Radioactivity Potentially Clearable from Various Facilities. May 30, 2001. Prepared for the Nuclear Regulatory Commission under contract NRC-04-01-049. Washington, D.C.: U.S. Nuclear Regulatory Commission.
Seiler, F., and J. Alvarez. 1996. On the Selection of Distributions for Stochastic Variables. Risk Analysis 16(1):5-18.
Sheppard, S.C., and W.G. Evenden (1997). Variation in Transfer Factors for Stochastic Models: Soil-to-Plant Transfer. Health Physics 72(5):727-733.
Shlyakhter, A. L., and Valverde A., Jr. 1995. Integrated Risk Analysis of Global Climate Change. Chemosphere 30(8): 1585-1618.
Smith, R. et al. 1996. Revised Analyses of Decommissioning for the Reference Boiling Water Reactor Power Station, NUREG/CR-6174 (July). Washington, D.C.: U.S. Nuclear Regulatory Commission.
Till, J. E., et al. 1995. The Utah Thyroid Cohort Study: Analysis of the Dosimetry Results. Health Physics 68(4):472-483.
United Nations Economic Commission for Europe (UNECE). 2001. Report on the Fourth Meeting of the Team of Specialists on Radioactive Contaminated Metal Scrap. Geneva: United Nations.
United Nations Scientific Committee on the Effects of Atomic Radiation (UNSCEAR). 1982. Ionizing Radiation: Sources and Biological Effects. The 1982 Report to the General Assembly with Annexes. New York: United Nations.

REFERENCES

UNSCEAR. 1988. Sources, Effects, and Risks of Ionizing Radiation. The 1988 Report to the General Assembly with Annexes. New York: United Nations.

UNSCEAR. 2000. Sources and Effects of Ionizing Radiation. Report to the General Assembly, with scientific annexes. New York: United Nations.

U.S. Army Corps of Engineers (USACE). 1998. United States Army Corps of Engineers. Radioactive Waste Disposal: Multiple Award Contracts for the Kansas City District, Request for Proposal. DACW41-99-0004, December. Kansas City, Kan.: USACE.

U.S. Nuclear Regulatory Commission (USNRC). 1980. Draft Environmental Statement Concerning Proposed Exemption from Licensing Requirements for Smelted Alloys Containing Residual Technetium-99 and Low-Enriched Uranium. NUREG-0518. October 1980. Washington, D.C.: USNRC.

USNRC. 1981. Control of Radioactively Contaminated Material. IE Circular 81-07. Office of Inspection and Enforcement. May 14. Washington, D.C.: USNRC.

USNRC. 1983. Guidelines for Decontamination of Facilities and Equipment Prior to Release for Unrestricted Use or Termination of Byproduct, Source or Special Nuclear Materials Licenses. Document FC 83-23. Washington, D.C.: USNRC.

USNRC. 1985. Surveys of Wastes Before Disposal from Nuclear Reactor Facilities. Information Notice 85-92. Washington, D.C.: USNRC.

USNRC. 1991a. Public Regional Meetings on Below Regulatory Concern Policy Statement August 28-September 27, 1990. SECY-91-087. April 1. Washington, D.C.: USNRC.

USNRC. 1991b. Evaluation of the Feasibility of Initiating a Consensus Process to Address Issues Related to the Below Regulatory Concern Policy. SECY-91-132. May 15, 1991. Washington, D.C.: USNRC.

USNRC. 1991c. Standards for Protection Against Radiation. Federal Register 56:23360-23474, May 21.

USNRC. 1991d. Memorandum dated June 28, 1991, from Samuel J. Chilk, secretary, USNRC, to James Taylor, executive director for operations, William Parler, general counsel, and Howard Denton, director of governmental and public affairs, USNRC.

USNRC. 1992. Enhanced Participatory Rulemaking Process. SECY-92-045. February 7. Washington, D.C.: USNRC.

USNRC. 1994. Draft Proposed Rule on Decommissioning. SECY 94-150. May 31. Washington, D.C.: USNRC.

USNRC. 1995. Probabilistic Accident Consequence Uncertainty Analysis: Dispersion and Deposition Uncertainty Assessment. Washington, Brussels: U.S. Nuclear Regulatory Commission and Commission of European Communities.

USNRC. 1997. Minimum Detectable Concentrations with Typical Radiation Survey Instruments for Various Contaminants and Field Conditions. Draft NUREG-1507. Washington, D.C.: USNRC.

USNRC. 1998a. Regulatory Options for Setting Standards on Clearance of Materials and Equipment Having Residual Radioactivity. SECY-98-028, June 30. Washington, D.C.: USNRC.

USNRC. 1998b. Radiological Assessments for Clearance of Equipment and Materials from Nuclear Facilities. Draft NUREG-1640. Washington, D.C.: USNRC.

USNRC. 1999a. Release of Solid Materials at Licensed Facilities: Issues Paper, Scoping Process for Environmental Issues, and Notice of Public Meetings. Federal Register 64: 35090–35100, June 30.

USNRC. 1999b. Letter from the Nuclear Regulatory Commission to the Honorable John Dingell, Committee on Commerce, U.S. House of Representatives. December 20.

USNRC. 1999c. Rulemaking Process in Response to the Staff Requirements Memorandum for SECY 98-028, Regulatory Options for Setting Standards on Clearance of Materials and Equipment Having Residual Radioactivity, January 27. Washington, D.C.: USNRC.

USNRC. 2000a. Control of Solid Materials: Results of Public Meetings, Status of Technical Analyses, and Recommendations for Proceeding. Policy Issue Information SECY-00-0070, March 23. Washington, D.C.: USNRC.

USNRC. 2000b. Report on Waste Burial Charges. NUREG-1307, Revision 9, September. Washington, D.C.: USNRC.

USNRC. 2000c. Meeting with Stakeholders on Efforts Regarding Release of Solid Material: Proceedings. Transcript of May 9, meeting.

USNRC. 2000d. Summary and Categorization of Public Comments on the Control of Solid Materials. NUREG/CR-6682. Report Prepared by ICF Consulting, Inc. September 2000. Washington, D.C.: USNRC.

USNRC. 2001a. Partial Response to SRM COMEXM-00-0002, Expansion of NRC Statutory Authority over Medical Use of Naturally Occurring and Accelerator-Produced Radioactive Material (NARM). Policy Issue Information SECY-01-0057 (March 29). Washington, D.C.: U.S. Nuclear Regulatory Commission.

USNRC. 2001b. Letter from Ashok Thadani, director, Office of Nuclear Regulatory Research, U.S. Nuclear Regulatory Commission, to Martin Offutt, program officer, National Research Council, written responses to the committee, April 16.

Wolbarst, A.B., et al. 1999. Sites in the United States Contaminated with Radioactivity.

Zuckerbrod, N. 2001. DOE Drops Nuclear Scrap Consultant (July 26). The Associated Press.

Appendixes

A

Biographical Sketches of Committee Members

Richard S. Magee, chair, is currently vice president, Carmagen Engineering, Inc., and technical director, New Jersey Corporation for Advanced Technology. His previous positions include associate provost for research and development and executive director, Otto H. York Center for Environmental Engineering and Science, New Jersey Institute of Technology (NJIT); professor in the Department of Mechanical Engineering and the Department of Chemical Engineering, Chemistry, and Environmental Science, NJIT; associate director, Environmental Protection Agency (EPA) Center for Airborne Organics, Massachusetts Institute of Technology (MIT); director, Northeast Hazardous Substance Research Center, NJIT; and director, Stevens Institute of Technology Energy Center. He has chaired numerous groups and committees including a number of National Research Council (NRC) committees. His NRC service includes chair and member of the NRC Evaluation Panel for the National Bureau of Standards Center for Fire Research; member of the Board of Assessment of the National Engineering Laboratories; chair and member of the Committee on Review and Evaluation of the Army Chemical Stockpile Disposal Program; chair and member of the Panel on Review and Evaluation of Alternative Chemical Disposal Technologies; and chair and member of the Committee on Review of the U.S. Department of Energy Office of Fossil Energy's Research Plan for Fine Particulates. He is a fellow of the American Society of Mechanical Engineers and a National Associate of the National Academies. He has also provided service to the North Atlantic Treaty Organization Science Committee as a member of the Priority Area Panel on Disarmament Technologies and as a member of the Advisory Panel on Security-related Civil Science and Technology. He has extensive experience in environmental science

183

and engineering, including expertise in combustion, incineration, emissions, hazardous waste, and energy technologies. He has a B.E., an M.S., and an Sc.D. from Stevens Institute of Technology.

David E. Adelman is an associate professor at the University of Arizona's James E. Rogers College of Law. His work focuses on the myriad interfaces between law and science, with particular emphasis on evaluating environmental and regulatory issues relating to new or controversial technologies as well as assessing the impacts of intellectual property regimes on scientific research in the United States. He is a member of the U.S. Department of Energy's (DOE's) Environmental Management Advisory Board, as well as a member of the National Academies' Committee on Building a Long-term Environmental Quality Research and Development Plan in the U.S. Department of Energy, which evaluated DOE's Environmental Management science program. From July 1998 to September 2001, he was a senior attorney with the Natural Resources Defense Council's (NRDC) Nuclear and Public Health programs in Washington, D.C., where he monitored and litigated issues pertaining to the environmental cleanup of the nuclear weapons complex and developed proposals for appropriate regulatory mechanisms for agricultural biotechnology. Prior to his position at NRDC, he was an associate at the law firm of Covington & Burling in Washington, D.C., where he litigated patent disputes and provided counsel on environmental regulatory issues. He received a B.A. in chemistry and physics from Reed College in 1988, a Ph.D. in chemical physics from Stanford University in 1993, and a J.D. from Stanford Law School in 1996.

Jan Beyea is a senior scientist with Consulting in the Public Interest and a consultant to the National Audubon Society and the Epidemiology Department of the Mount Sinai Medical School. He consults on nuclear physics and other energy and environmental topics for numerous local, national, and international organizations. He has been chief scientist and vice president, National Audubon Society, and has held positions at the Center for Energy and Environmental Studies, Princeton University, Holy Cross College, and Columbia University. He has served on numerous advisory committees and panels including as a member of the NRC's Board on Energy and Environmental Systems, Energy Engineering Board; Committee on Alternative Energy R&D Strategies; and Committee to Review DOE's Fine Particulates Research Plan. He has served on the Secretary of Energy's Advisory Board, Task Force on Economic Modeling and the policy committee of the Recycling Advisory Council. He served as an advisor to various Office of Technology Assessment studies. He has expertise in energy technologies and associated environmental and health concerns and has written numerous articles on environment and energy. He received a B.A. from Amherst College and a Ph.D. in physics from Columbia University.

Jack S. Brenizer, Jr., is a professor in the Department of Mechanical and Nuclear Engineering and chairman of the Nuclear Engineering Program at the Pennsylvania State University. His previous positions include associate professor, School of Engineering and Applied Science, University of Virginia, Charlottesville, and engineering technician, AMP Incorporated. His research and teaching interests cover a wide range of expertise related to nuclear science and engineering, nuclear measurements, radiation detection, reactor operations and systems, and effects of radiation. He is a recipient of the American Society for Testing and Materials (ASTM) E7 Charles W. Briggs Award and a Board Member of the International Society for Neutron Radiography. He is a member of the American Nuclear Society, the Health Physics Society, Sigma Xi, the American Society for Nondestructive Testing, the ASTM, the International Society for Neutron Radiography, and the International Society for Optical Engineering. He has a B.S. in physics from Shippensburg State College and an M.E. (engineering science) and a Ph.D. (nuclear engineering) from the Pennsylvania State University.

Lynda L. Brothers is a partner with Sonnenschein Nath & Rosenthal. Her previous positions include partner, Davis Wright Tremaine (1990-2000); executive vice president, Raytheon Hanford, Inc. (1996); assistant director, Hazardous, Solid and Radioactive Waste and Air Quality, Department of Ecology, State of Washington (1983-1985); deputy assistant secretary for environment, U.S. Department of Energy (1979-1981); and counsel, Subcommittee on Environment and Atmosphere, Committee on Science and Technology, U.S. House of Representatives (1978-1979). She has extensive experience in environmental and radioactive waste issues that cut across many agencies and jurisdictions and addresses regulatory issues related to defense wastes and commercial low-level radioactive waste. She has served on a number of advisory boards and committees including the NRC's Board on Radioactive Waste Management (1989-1996); Committee on Classification of Documents at the Department of Energy; Committee to Review New York State's Siting and Methodology Selection for Low Level Radioactive Waste Disposal. She has also served on the Advisory Board, Virginia Mason Center for Women's Health; the Northwest Citizens' Forum on High Level Nuclear Waste at Hanford; and the Board of Trustees, Washington Environmental Foundation. She served as chair of the Northwest Interstate Compact Commission on Low Level Radioactive Waste from 1983-1985. Until spring of 2000, she was counsel to the board of directors of Envirocare of Utah, and she currently serves on the board of directors, American Birding Association. She has a J.D. from the Golden State University, an M.S. in biology from the University of Virginia, and a B.S. in genetics from the University of California, Berkeley.

Robert J. Budnitz is president of Future Resources Associates, Inc., in Berkeley, California. Previously, he served as deputy director and director of the U.S.

Nuclear Regulatory Commission's (USNRC's) Office of Nuclear Regulatory Research and also held several management positions at the Lawrence Berkeley Laboratory of the University of California. His professional interests are in environmental impacts, hazards, and safety analysis, particularly of the nuclear fuel cycle. He has been prominent in the field of nuclear reactor safety assessment and waste repository performance assessment, including probabilistic risk assessment. Dr. Budnitz has served on numerous investigative and advisory panels of scientific societies, government agencies, and the National Research Council. His most recent NRC committee service was with the Board on Radioactive Waste Management, Committee on Buried and Tank Wastes and Committee on Technical Bases for Yucca Mountain Standards. He is a member of the Board of Directors of the Cal Rad Forum, an association of public and private institutions and corporations that generate low-level radioactive waste in the Southwestern Low-Level Waste Disposal Compact, which supports the prompt development of the Ward Valley site in California. He received a B.A. from Yale University and a Ph.D. in physics from Harvard University.

Gregory R. Choppin is currently the R.O. Lawton Distinguished Professor of Chemistry at Florida State University. His research interests involve the chemistry of the f-elements, the separation science of the f-elements, and the physical chemistry of concentrated electrolyte solutions. During a postdoctoral period at the Lawrence Radiation Laboratory, University of California, Berkeley, he participated in the discovery of mendelevium, element 101. His research and educational activities have been recognized by the American Chemical Society's Award in Nuclear Chemistry, the Southern Chemist Award of the American Chemical Society, the Manufacturing Chemist Award in Chemical Education, the Chemical Pioneer Award of the American Institute of Chemistry, a Presidential Citation Award of the American Nuclear Society, and honorary D.Sc. degrees from Loyola University and the Chalmers University of Technology (Sweden). Dr. Choppin has served as member, chair, or vice chair of numerous NRC committees and is currently a member of the Board on Radioactive Waste Management and chair of the Committee on Building a Long-term Environmental Quality Research and Development Program in the U.S. Department of Energy. Dr. Choppin received a B.S. in chemistry from Loyola University, New Orleans, and a Ph.D. from the University of Texas, Austin.

Michael Corradini (National Academy of Engineering [NAE]) is a professor in the Department of Engineering Physics at the University of Wisconsin, Madison, and associate dean of the College of Engineering. Dr. Corradini's research focus is nuclear engineering and multiphase flow with specific interests that include light-water reactor safety, fusion reactor design and safety, waste management and disposal, vapor explosions research and molten core concrete interaction research, and energy policy analysis. He is a member of the American Institute of

Chemical Engineers, the American Society of Engineering Education, and the American Society of Mechanical Engineers, and a fellow of the American Nuclear Society. Dr. Corradini has received numerous awards including the National Science Foundation's Presidential Young Investigators Award, the American Nuclear Society's reactor safety best paper award, and the University of Wisconsin, Madison campus, teaching award. He is the author of more than 100 technical papers and has served on various technical review committees, including the research review panel of the USNRC and the direct heating review group. He is currently a member of the NRC's Electric Power/Energy Systems Engineering Peer Committee and chair of the Frontiers of Engineering Organizing Committee. Dr. Corradini was elected to the NAE in 1998. He received his B.S. in mechanical engineering from Marquette University and his M.S. and Ph.D. in nuclear engineering from the Massachusetts Institute of Technology.

James W. Dally (NAE) is Glenn L. Martin Institute Professor of Engineering Emeritus, University of Maryland, College Park. Dr. Dally has had a distinguished career in industry, government, and academia and is the former dean of the College of Engineering at the University of Rhode Island. His former positions include senior research engineer, Armour Research Foundation; assistant director of research, Illinois Institute of Technology Research Institute; assistant professor, Cornell University; professor, Illinois Institute of Technology; and senior engineer, International Business Machines Corporation. He is also an independent consultant. Dr. Dally is a mechanical engineer and the author or coauthor of six books, including engineering textbooks on experimental stress analysis, engineering design, instrumentation, and the packaging of electronic systems, and has published approximately 200 research papers. He has served on a number of NRC committees and is currently on the Committee on the Future Environments for the National Institute for Standards and Technology and the Committee on Review of Federal Motor Carrier Safety Administration's Truck Crash Causation Study. He has a B.S. and an M.S. from the Carnegie Institute of Technology, and a Ph.D. from the Illinois Institute of Technology.

Edward R. Epp is professor of radiation oncology, emeritus, Harvard University. He has served as physicist, Department of Radiology, Montreal General Hospital; has worked at the Sloan-Kettering Institute for Cancer Research where he served as member and professor of biophysics at Cornell University in the Graduate School of Medical Sciences; was professor of radiation oncology, Harvard Medical School; and served as head of the Division of Radiation Biophysics in the Department of Radiation Oncology at Massachusetts General Hospital. Dr. Epp is a fellow of the American Physical Society and the American Association of Physicists in Medicine. He has served as president of the Radiation Research Society and on a number of committees of the National Academy of Sciences. He has also been a member of the National Institute of Health

Radiation Study Section and the National Cancer Institute Clinical Program Project Committee. In 2000, he was the Failla Memorial Lecturer for the Greater New York Chapter of the Health Physics Society in association with the Radiation and Medical Physics Society of New York. His research interests include radiation physics and dosimetry, radiation biophysics, and mechanisms of radiation action in cells. His specific research on mechanism aspects has dealt with the biological effects of ultrahigh-intensity pulsed radiation in the presence of oxygen and other chemical sensitizers. He obtained his B.A. and M.A. degrees from the University of Saskatchewan and his Ph.D. in nuclear physics from McGill University.

Alvin Mushkatel is currently a professor in the School of Planning and Landscape Architecture at Arizona State University (ASU). Previous positions at ASU include professor, School of Public Affairs; director of the Doctor of Public Administration Program; and director of the Office of Hazards Studies. He has held positions in political science at the University of Denver; University of Missouri, St. Louis; and St. John's University in Minnesota. He has conducted numerous studies and published widely in a number of areas including risk perception, siting of hazardous waste facilities, public and stakeholder involvement in policy making, and nuclear waste policy. He has served on numerous advisory bodies and committees including the U.S. Department of Energy Headquarters Public Participation Seminar Series Panel on public trust and confidence, and on the following NRC committees: Earthquake Engineering and a number of its subpanels; Committee on Review and Evaluation of the Army Chemical Stockpile Disposal Program; Committee to Assess the Policies and Practices of the Department of Energy to Design, Manage, and Procure Environmental Restoration, Waste Management, and Other Construction Projects; and Committee on Decontamination and Decommissioning of Uranium Enrichment Facilities. Dr. Mushkatel received his Ph.D. in political science from the University of Oregon.

Rebecca R. Rubin is a partner in the BAHR Environmental Company, in which she leads and performs environmental studies and evaluations for clients in the federal and commercial sectors. She has held a number of positions in the environmental field including director, Army Environmental Policy Institute, managing the research, analysis, and development of progressive environmental policies and strategies for the U.S. Army; and manager, project leader, and analyst, Environmental Program, Institute for Defense Analyses, where she managed the environmental studies program and conducted studies for the Department of Defense and other government agencies. Her experience in the environmental area covers a broad range of subjects including the integration of environmental, safety, and health considerations with defense acquisition; evaluation of site contamination, developmental testing of environmental technologies; and policies and strategies for environmental cleanup and compliance. She has a B.A. from Harvard College and an M.A. from Columbia University.

Michael T. Ryan is an associate professor, Department of Health Administration and Policy, Medical University of South Carolina (MUSC). He earned his B.S. in radiological health physics from Lowell Technological Institute in 1974. In 1976, he earned his M.S. in radiological sciences and protection from the University of Lowell. Dr. Ryan received a Ph.D. in 1982 from the Georgia Institute of Technology, where he was recently inducted into the Academy of Distinguished Alumni. Dr. Ryan is an editor in chief of *Health Physics Journal*. Over the past 10 years, he has served on the Technical Advisory Radiation Control Council for the State of South Carolina. He is a member of the National Council of Radiation Protection and Measurements (NCRP) scientific vice president for Radioactive and Mixed Waste Management and chair of Scientific Committee 87; and a member of the board of directors. He is also a member of NCRP's scientific committee 87-4 on Management of Waste Metals Containing Radioactivity. Dr. Ryan is certified in the comprehensive practice of health physics by the American Board of Health Physics. Dr. Ryan holds adjunct appointments at Georgia Tech and at the University of South Carolina and the College of Charleston where he has taught radiation protection courses at the graduate level. He is currently serving on the Scientific Review Group appointed by the Assistant Secretary of Energy to review the ongoing research in health effects at the former weapons complex at Mayak in the Southern Urals of the former Soviet Union. Prior to his appointment at MUSC, Dr. Ryan was most recently vice president of Barnwell Operations for Chem-Nuclear Systems, Inc., and previously served as vice president of regulatory affairs, having responsibility for developing and implementing the company's policies and programs to comply with state and federal regulations. Before joining Chem-Nuclear Systems, Inc., as director of the Environmental and Dosimetry Laboratory in 1983, Dr. Ryan spent seven years in environmental health physics research at Oak Ridge National Laboratory.

Richard I. Smith retired from Pacific Northwest National Laboratory in 1996 after nearly 40 years of scientific activities on the Hanford Site, where he was a staff engineer in the Systems and Risk Management Department. He has extensive experience related to decontamination and decommissioning (D&D) of licensed nuclear facilities, including cost analyses and environmental impact analyses. His studies on the decommissioning of power and test reactors, fuel cycle facilities, and non-fuel cycle nuclear facilities, which focus on estimating the costs and occupational radiation dose for D&D of nuclear facilities, are known and used throughout the world. He has participated in the development of several reports for the International Atomic Energy Agency (IAEA) on the D&D of nuclear facilities, dealing with the status of technology decontamination, disassembly, and waste management, and he served as a member of an IAEA working group considering the planning for decommissioning of WWER-440 reactors throughout the former Eastern bloc countries. He has also recently contributed to the International Nuclear Safety Program in the area of planning for decommissioning the three undamaged reactors at the Chernobyl Nuclear Power Station in Ukraine. He has led studies in the storage, packaging, and transport of spent fuel and greater than Class C waste. He has served on the NRC Committee on Decontamination and Decommissioning of Uranium Enrichment Facilities, and the Committee to Assess the Policies and Practices of the DOE to Design, Manage and Procure Environmental Restoration, Waste Management, and Other Construction Projects. He has a B.S. in physics from Washington State University and an M.S. in applied physics from the University of California at Los Angeles; he is a professional engineer in nuclear engineering, licensed in the states of Washington and California.

Dale Stein (NAE) is president emeritus of Michigan Technological University and retired professor of materials science. He has held positions at Michigan Technological University, the University of Minnesota, and the General Electric Research Laboratory. He is a recipient of the Hardy Gold Medal of the American Institute of Mining, Metallurgical and Petroleum Engineers and the Geisler Award of the American Society of Metals (Eastern New York Chapter), and he has been an elected fellow of the American Society of Metals and the American Association for the Advancement of Science. He has served on numerous NRC committees: he is currently a member of the Committee on Review of DOE's Office of Heavy Vehicle Technologies; Committee on Review of National Transportation Science and Technology Strategy; and Research and Technology Coordinating Committee of the Transportation Research Board; he was chair of the Committee on Decontamination and Decommissioning of Uranium Enrichment Facilities. He previously was a member of the U.S. Department of Energy's Energy Research Advisory Board. He is currently chairman of the Advisory Committee for the Center for Nuclear Waste Regulatory Analyses (CNWRA), which is concerned primarily with advising the USNRC on the granting of a license for a repository for high-level nuclear waste; CNWRA is affiliated with the Southwest Research Institute, a contractor to the USNRC. He is also a member of the NAE and is an internationally known authority on the mechanical properties of engineering materials. He received his Ph.D. in metallurgy from Rensselaer Polytechnic Institute.

Detlof von Winterfeldt is a professor of public policy and management at the University of Southern California and director of its Institute for Civic Enterprise. He also is the president of Decision Insights, Inc., a management consulting firm specializing in decision and risk analysis. His research interests are in the foundation and practice of decision and risk analysis as applied to technology and environmental management problems. He is the coauthor of two books and author or coauthor of more than 100 articles and reports on these topics. He has served on several committees and panels of the National Science Foundation (NSF) and the National Research Council, including the NSF's Advisory Panel for its Decision and Risk Management Science Program and the NRC's Committee on Risk Perception and Risk Communication.

B

Presentations and Committee Activities

1. **Committee Meeting, National Academy of Sciences, Washington, D.C., January 3-5, 2001**

    Controlling the Release of Solid Materials
    Richard A. Meserve, Chairman, U.S. Nuclear Regulatory Commission

    EPA's Clean Materials Program
    Craig Conklin, Office of Air and Radiation, U.S. Environmental Protection Agency

    Revision of DOE Requirements for Control of the Release of Materials for Re-use and Recycle
    Andrew Wallo, Office of Environmental Safety and Health, U.S. Department of Energy

    Controlling Release of Solid Materials—Current Approach
    Anthony Huffert, U.S. Nuclear Regulatory Commission

    Controlling Release of Solid Materials—Public Input
    Frank Cardile, U.S. Nuclear Regulatory Commission

    Controlling Release of Solid Materials—International Status
    Robert Meck, U.S. Nuclear Regulatory Commission

    Controlling Release of Solid Materials—Technical Bases
    Robert Meck, U.S. Nuclear Regulatory Commission

2. **Committee Meeting, National Academy of Sciences, Washington, D.C., March 26-28, 2001**

Radiological Clearance: An Industry Perspective
Paul Genoa, Nuclear Energy Institute

NAS—Release of Radioactive Material
George Vanderheyden, Exelon

Maine Yankee Atomic Power Company: Decommissioning Update
William O'Dell, Entergy Corporation

Big Rock Point Restoration Project
Kurt Haas, Consumers Energy

Release of Solid Materials
Ellen Heath, Duke Engineering

Envirocare of Utah, Inc.: The Safe Alternative
Charles Judd, Envirocare

Vehicle Radiation Monitoring Systems
Jas Devgun, American Nuclear Society

Presentation to National Research Council Committee on Alternatives for Controlling the Release of Solid Materials from NRC-Licensed Facilities
Gary Visscher, American Iron and Steel Institute

Radiation and Steel
Anthony LaMastra, Health Physics Associates

Restricted Recycling of Metals
Gordon Geiger, University of Arizona

National Academy of Sciences Presentation
Eric Stuart, Steel Manufacturers Association

Washington State's Perspective on Controlling the Release of Solid Materials from Nuclear Facilities
John Erickson, State of Washington, Division of Radiation Protection

Presentation to National Academy of Sciences
Henry Porter, S.C. Department of Health and Environmental Control

Comments on Clearance Rules
John Erickson, Organization of Agreement States

Position Statement of CRCPD
Kathleen McAllister, Conference of Radiation Control Program Directors

Radioactivity in Solid Waste
David Allard, State of Pennsylvania Bureau of Radiation Protection

Federal Solid Waste Disposal Regulations
Bob Dellinger, US Environmental Protection Agency, Office of Solid Waste

Radioactive Materials Found in Municipal Waste and Recycle Materials
Greg Smith, Radiation Service Organization

Statement to the Committee
Dan Guttman, Attorney-at-law

Committee Must Safeguard Public Health and Allow More Public Interest Input
David Ritter, Public Citizen

Radioactive Waste and Materials Release and Recycling
Diane D'Arrigo, Nuclear Information and Resource Service

Presentation to Committee on Alternatives for Controlling the Release of Solid Materials from NRC-Licensed Facilities
Jens Hovgaard, Exploranium G.S. Ltd.

Radioactive Waste Management at Stanford Linear Accelerator
Steven Frey, Stanford Linear Accelerator Center

Brokering, Assaying, and Releasing "Potentially Clean" Waste
Al Johnson, Duratek, Inc.

Demolition Waste and Metals Recycling
Al Johnson, Duratek, Inc.

Comments of the National Ready Mixed Concrete Association
Robert Garbini, National Ready Mixed Concrete Association

3. Committee Subgroup Site Visit to ATG, Richland, Washington, April 16, 2001

4. Committee Subgroup Site Visit to Duratek Inc., Oak Ridge, Tennessee, June 1, 2001

5. Committee Meeting, National Academy of Sciences, Washington, D.C., June 12-15, 2001

Discussion of EPA Technical Support Document
Robert Anigstein, Sanford Cohen & Associates

Discussion of NUREG-1640
Robert Meck, U.S. Nuclear Regulatory Commission

Recyclable Metallurgical Scrap Metal for the Steel Industry
Ray Turner, Health Physicist, David J. Joseph Company

Scope of International Regulations
Gordon Linsley, Waste Safety Section, International Atomic Energy Agency

Application of the Concepts of Clearance in the European Union
Augustin Janssens, Environment Directorate-General, European Commission

A Nuclear Decommissioner's Views on Clearance Levels
Shankar Menon, OECD/NEA Cooperative Programme on Decommissioning

Stakeholder Involvement Strategies for Highly Technical and Controversial Issues
Janesse Brewer, Senior Facilitator, The Keystone Center

SAIC Organizational Conflict of Interest
Dan Guttman, Attorney-at-Law

A Historical Perspective on the NRC Public Participation Process After the BRC Policy
Francis Cameron, Special Counsel for Public Liaison, U.S. Nuclear Regulatory Commission

6. **Committee Subgroup Meeting, National Academy of Sciences, Washington, D.C., July 16-18, 2001**

7. **Committee Subgroup Meeting, National Academy of Sciences, Washington, D.C., July 30-August 1, 2001**

8. **Committee Meeting, Woods Hole, Massachusetts, August 29-31, 2001**

9. **Committee Meeting, National Academy of Sciences, Washington, D.C., November 19-20, 2001**

C

Statement of Work

The National Research Council committee formed to undertake this study will address the following tasks:

(1) As part of its data gathering and understanding the technical basis for the Nuclear Regulatory Commission's (USNRC's) analyses of various alternatives for managing solid materials from USNRC-licensed facilities, the committee shall review the technical bases and policies and precedents derived therefrom set by the USNRC and other Federal agencies, by States, other nations and international agencies, and other standard setting bodies including the following. The review of the following will be contingent on the USNRC staff providing summaries with the salient issues of each document to the Research Council staff and committee, as well as copies of the documents, soon after project funds are received and before the first committee meeting.

- The USNRC technical bases development, including ongoing and planned staff activities, to include the assessment of potential scenarios and pathways for radiation exposure, survey and detection methodology, and an evaluation of the environmental impacts for a variety of solid materials.
- The 1997 Environmental Protection Agency Preliminary Technical Support Document for its clean metals program and other studies on the environmental impacts of clearance of materials, exemption of materials containing naturally occurring radioactive material (e.g., coal ash), and development of guidelines for screening materials imported into the U.S. that contain radioactivity.
- The 1980 Department of Energy (DOE) petition to establish exemptions

APPENDIX C   197

for small concentrations of technetium-99 and/or low enriched uranium as residual contamination in smelted alloys and the public comment on the proposed DOE rule.
- The 1990 USNRC Below Regulatory Concern (BRC) Policy setting a standard for release of solid materials for recycle. In 1991 the USNRC instituted a moratorium on the BRC Policy to allow more extensive public involvement, and the BRC policy was revoked by Congress in the Energy Policy Act of 1992.
- DOE criteria (e.g., DOE Order 5400.5) for release of solid materials and handbooks for controlling release of property containing residual radioactive material. DOE has established a task force to review its policies on release of materials for re-use and recycling that could have implications for USNRC licensees.
- Conference of Radiation Control Program Directors recommendations or policies on the control of solid materials from licensed facilities.
- Experience of individual states promulgating release criteria for solid materials in the absence of federal standards. For example, one state prohibits the disposal of radioactive material in municipal landfills and another state authorizes unrestricted release of volumetrically contaminated materials. Methodologies states are using to survey and detect slightly contaminated materials. Basis and criteria states are using for approving the release of these materials. Approaches states are using for similar levels of naturally occurring radioactive materials.
- International Atomic Energy Agency and European Union experience, directives, recommendations or standards, especially as they pertain to international adoption of guidelines and criteria on international trade and import standards.
- Recommendations of the International Commission on Radiological Protection (e.g., ICRP Report 60) and the National Council on Radiation Protection and Measurements (e.g., NCRP Report 116) and on-going activities evaluating clearance and criteria for release of slightly radioactive materials.
- American National Standards Institute Standard N13.12, "Surface and Volume Radioactivity Standards for Clearance." This standard contains criteria for unrestricted release of solid materials from nuclear facilities. Also, review of the National Technology Transfer and Advancement Act of 1995 and its implications for developing and implementing alternative release criteria.

(2) The committee will review public comments and reactions received so far on current and former USNRC proposals to develop alternatives for control of solid materials. Again, this review will be contingent on the USNRC staff providing the committee both with the comments and summaries of the public com-

ments and reactions received. The committee will explicitly consider how to address public perception of risks associated with the direct reuse, recycle, or disposal of solid materials released from USNRC-licensed facilities. The committee should provide recommendations for USNRC consideration on how comments and concerns of stakeholders can be integrated into an acceptable approach for proceeding to address the release of solid materials.

(3) The committee shall determine whether there are sufficient technical bases to establish criteria for controlling the release of slightly contaminated solid materials. This should include an evaluation of methods to identify the critical groups, exposure pathway(s), assessment of individual and collective dose, exposure scenarios, and the validation and verification of exposure criteria for regulatory purposes (i.e., decision making and compliance). As part of this determination, it should judge whether there is adequate, affordable measurement technology for USNRC-licensees to verify and demonstrate compliance with a release criteria. What, if any, additional analyses or technical bases are needed before release criteria can be established?

(4) Based on its evaluation and its review, the committee shall recommend whether USNRC: (1) continue the current system of case-by-case decisions on control of material using existing, revised, or new (to address volumetrically contaminated materials) regulatory guidance, (2) establish a national standard by rulemaking, to establish generic criteria for controlling the release of solid materials, or (3) consider another alternative approach(es).

If the committee recommends continuation of the current system of case-by-case decisions, the committee shall provide recommendations on if and how the current system of authorizing the release of solid materials should be revised.
If the committee recommends that USNRC promulgate a national standard for the release of solid material, the committee shall: (1) recommend an approach, (2) set the basis for release criteria (e.g., dose, activity, or detectability-based), and (3) suggest a basis for establishing a numerical limit(s) with regard to the release criteria or, if it deems appropriate, propose a numerical limit.

(5) The committee shall make recommendations on how the USNRC might consider international clearance (i.e., solid material release) standards in its implementation of the recommended technical approach.

D

Standards (Limits) Proposed by Other Organizations

AMERICAN NATIONAL STANDARDS INSTITUTE AND HEALTH PHYSICS SOCIETY

ANSI/HPS N13.12-1999 Surface and Volume Radioactivity Standards for Clearance

The Health Physics Society (HPS) Standards Working Group developed this standard. The standard was consensus balloted[1] and approved by the American National Standards Institute (ANSI) accredited HPS N13 Committee on October 19, 1998. Furthermore, ANSI, Inc., itself approved the standard on August 31, 1999. The standard defines primary (dose) and secondary screening (derived) criteria.

Primary Dose Criterion

The primary criterion of this standard is to provide for the public health and safety of an average member of a critical group such that the dose shall be limited to 10 μSv/yr (1.0 mrem/yr) total effective dose equivalent (TEDE), above background, for clearance of materials from regulatory control. When justified on a case-by-case basis, clearance shall be permitted at higher dose levels when it can be ensured that exposures to multiple sources will be maintained as low as rea-

---

[1] A listing of the organizations and government agencies represented on the N13 Committee is listed in an Appendix to the ANSI/HPS standard.

sonably achievable (ALARA) and will provide an adequate margin of safety below the public dose limit of 1 mSv/yr (100 mrem/yr) TEDE.

Derived Screening Levels

Derived screening levels, above background, for the clearance of solid materials or items containing surface or volume activity concentrations of radioactive materials are tabulated in the standard. In that table the radionuclides have been divided into four groups based on similarity of exposure scenario results, with screening levels ranging from 0.1 to 100 Bq/cm^2 (or Bq/g), depending on the group considered.[2] A generic ALARA process was employed in developing the derived screening levels. However, based on a detailed ALARA evaluation, it shall be permissible to derive less restrictive screening levels on a case-by-case basis using the primary dose criterion.

The standard includes a discussion of the collective dose in relation to the screening levels. In reality, concentrations in cleared materials will likely average about an order of magnitude less than the screening levels, which are intended to define upper bounds. From the qualitative evaluation it is concluded that on the average, individuals will likely receive no more than the 10 μSv/yr (1.0 mrem/yr) primary dose criterion because of conservative modeling and assumed maximum concentrations. Assuming an average U.S. background level of 3.0 mSv/yr (300 mrem/yr), the collective doses to the critical group resulting from clearance of items using the criterion from this standard will be no more than 0.3 percent of the dose the same population would receive from natural background radiation in any one year. The magnitude of the potential collective doses to the critical group associated with the items in accordance with this standard is so low that additional ALARA evaluations or analyses, or further reductions in the primary dose standard, are not deemed necessary.

INTERNATIONAL ATOMIC ENERGY AGENCY

Safety Series No. 89: Principles for the Exemption of Radioactive Sources and Practices from Regulatory Control

This document was jointly sponsored by the International Atomic Energy Agency (IAEA) and the Nuclear Energy Agency of the Organization for Economic Cooperation and Development and was published in 1988. It is based on two principles for exemption:

---

[2]*Surface and Volume Radioactivity Standards for Clearance: An American National Standard*, Health Physics Society Report, ANSI/HPS N13.12-1999.

1. Individual risk must be sufficiently low as not to warrant regulatory concern.
2. Radiation protection, including the cost of regulatory control, must be optimized.

Two approaches were followed in determining if the level of risk or dose is trivial;[3] first, choose a level of risk and the corresponding dose that is of no significance to individuals; second, use the exposure to natural radiation, to the extent that it is normal and unavoidable, as a relevant reference level.

Risk-Based Considerations

It is widely recognized that values of individual risk that can be treated as insignificant correspond to a level at which individuals, aware of these risks, would not commit significant resources of their own to reduce them. It is believed that few people would commit their own resources to reduce an annual risk of death of 10^{-5} and that even fewer would take action at an annual level of 10^{-6}. By considering a rounded risk factor of 10^{-2} Sv^{-1} (10^{-4} rem^{-1}) for whole-body exposure as a broad average over age and gender, the level of trivial individual effective dose equivalent would be in the range of 10 to 100 mSv/yr (1 to 10 mrem/yr).[4]

Natural Background Radiation Considerations

The natural background radiation has been estimated to give an average individual dose of about 2.4 mSv/yr (240 mrem/yr).[5] This average conceals a wide variation due to different concentrations of radioactive materials in the ground and in building materials, different altitudes, and different habits of people. About half of this dose is due to radon exposure, which may be controlled. The other half comes from cosmic rays, terrestrial gamma rays, and radionuclides in the body for which control is not practical. Individuals do not usually consider variation in exposure to natural background radiation when considering moving

---

[3]The word trivial is used extensively by the IAEA in Vienna and the European Commission in their safety guides when describing an individual effective dose equivalent in the range of 10 to 100 mSv/yr (1 to 10 mrem/yr).

[4]*Principles for the Exemption of Radiation Sources and Practices from Regulatory Control*, Safety Series No. 89, International Atomic Energy Agency, Vienna, 1988.

[5]The background radiation varies significantly from country to country and from one location to another within a country. There are several regions in the world where natural background radiation gives doses that exceed the normal ranges by factors of 4 to 6. It is reported that no adverse health effects have been discerned from doses arising from these high natural levels. See, BEIR V, National Academy Press, Washington, D.C., 1990.

from one location to another or when going on a holiday. IAEA believes it can therefore be judged that a dose level that is small in comparison with the variation in natural background radiation can be considered trivial. A figure of whole-body or effective dose equivalent of the order of one to a few percent of the natural background, 20 to 100 mSv/yr (2 to 10 mrem/yr), has been suggested.

Conclusion on Individual Related Risk

The IAEA concluded that an individual radiation dose, regardless of its origin, is likely to be trivial if it is of the order of some tens of microsieverts per year (some millirems per year). It is noted that this dose corresponds to a few percent of the annual dose limit for members of the public recommended by the International Commission on Radiological Protection (ICRP) in 1977 and is much lower than any upper bound set by competent authorities for practices subject to regulatory control.

EUROPEAN COMMISSION

Radiation Protection 89: Recommended Radiological Protection Criteria for the Recycling of Metals from the Dismantling of Nuclear Installations

This document provides recommended radiological protection criteria for the recycling of metal arising from dismantling nuclear installations. The document prepared by the Group of Experts established under the terms of Article 31 of the Euratom Treaty confirms and extends its previous recommendations, made in IAEA Safety Series 89 of 1988. It has been demonstrated that metals below the clearance levels specified can be released from regulatory control with negligible risk, from a radiation point of view, for workers in the metals industry and for the population at large.

Radiological Protection Criteria

The document references the IAEA recommendation in Safety Series 89 that an individual dose of some tens of microsieverts is considered trivial and therefore the basis for exemption. The Working Group adopted radiation protection levels of 10 µSv/yr (1 mrem/yr) and 1 man-Sv/yr (100 man-rem/yr) of practice collective dose criteria. In addition, the skin dose was limited to 50 mSv/yr (5 rem/yr).

Relating the dose received by individuals to a practice, and to the levels of radioactivity involved in a practice, is difficult because the clearance criteria must be defined for a largely hypothetical environment. The Working Group constructed a set of exposure scenarios, which relate the activity content of the recycled metals to an individual dose. The proposed clearance levels are derived radioactivity levels from the most critical scenario.

Tables are provided that specify clearance levels for metal scrap recycling for the radionuclides encountered in decommissioning. A similar table is provided specifying the more stringent clearance levels for direct reuse of the metal.

AMERICAN NUCLEAR SOCIETY

The American Nuclear Society (ANS) Special Committee on Site Cleanup and Restoration Standards is responsible for reviewing draft regulations from federal organizations related to the decommissioning of nuclear facilities and providing ANS input to the rulemaking process.

The ANS is not in the business of writing standards, although it does write position papers and makes comments after reviewing rules. As of this date, the ANS has not endorsed ANSI N13.12. However, in a letter of December 4, 2000, ANS made the following comments regarding the Department of Energy's (DOE's) standard:

- ANS considers the 1 mrem/yr standard to be unreasonably low and without a firm scientific justification.
- Scientific evidence would seem to support a dose limit several times larger than the proposed 1 mrem/yr.

The ANS is currently working on a draft position paper on the standard for clearance and expects it to be released in 2002.

INTERNATIONAL COMMISSION ON RADIOLOGICAL PROTECTION

Publication 60

The ICRP recommends that the maximum permissible dose for occupational exposure should be 20 mSv/yr (2,000 mrem/yr), averaged over five years (i.e., 100 mSv total) with a maximum of 50 mSv in any one year. For public exposure, 1 mSv/yr (100 mrem/yr), averaged over 5 years, is the limit. In both categories, the figures are over and above background levels and exclude medical exposure.

The ICRP proposed apportionment of the total allowable dose from all anthropogenic sources of radiation (excluding medical exposures). Hence for radioactive waste management, authorities could allocate a fraction of the 1 mSv/yr (100 mrem/yr), to establish an exposure limit for low-level radioactive waste (LLRW). For example, the Environmental Protection Agency (EPA) in 40 CFR Part 191 imposed a limit of 0.15 mSv/yr (15 mrem/yr), which is consistent with the ICRP's concept of apportionment.

THE EUROPEAN UNION

Basic Safety Standards

The scope of the Basic Safety Standards (BSS) adopted by the European Union (EU) is defined in terms of practices and only indirectly in terms of radioactive substances. Justification of any practice involving radioactivity is required, i.e., determining whether the benefits to individuals and to society from introducing or continuing the practice outweigh the harm (including radiation detriment) resulting from the practice. If such practice is deemed justifiable, a decision is made as to whether it should be placed under the system of reporting and prior authorization as described by the BSS. Exempt practices are those that do not fall under this system. Practices thought to involve appreciable potential risks are put under the regulatory system without exception, including all of the practices associated with the nuclear fuel cycle. However, practices can be exempt from control if the associated risks are sufficiently low. Exemption levels have been derived for the BSS that apply to radionuclide levels and activities per unit mass from which the risks are regarded as trivial.

All associated activities and material movements are regulated after a practice has been placed within the regulatory system. Regulatory control can be relinquished only by proceeding through the system of reporting and prior authorization set out by the BSS. An ad hoc case-by-case procedure is followed for the possible release of materials for recycling, reuse, and disposal, and implementation of this procedure is the responsibility of the competent national authorities. Clearance is defined as the removal from regulatory control of a substance that has radionuclide levels below the recommended specific limits. Cleared substances are automatically exempt from the requirements of reporting and authorization.

The radiological protection criteria that have been adopted for clearance are 10 μSv/yr (1 mrem/yr), with a collective dose for the practice of 1 man-Sv/yr (100 man-rem/yr).[6] Additionally, the skin dose is limited to 50 μSv/yr.

European Union Directive 96/29/EURATOM allows clearance of "radioactive substances where the concentration of activity per unit mass do not exceed the exemption values set out in Column 3 of Table A to Annex I." Annex I is reproduced at the end of this appendix (see Table D-1), as is a table of derived clearance levels based on a primary dose standard of 10 μSv/yr from NUREG-1640 (Table D-2).

---

[6]European Union Directive 96/29/EURATOM further provides that collective dose can exceed 1 man-sievert, provided that "an assessment of an optimization of protection shows that exemption is the optimum option" (EU, 1996, p. 19).

UNITED NATIONS

United Nations Scientific Committee on the Effects of Atomic Radiation (UNSCEAR)

Consistent with the current understanding of the related consequences, the ICRP, National Council on Radiation Protection and Measurements (NCRP), IAEA, and UNSCEAR have recommended that radiation doses above background levels to members of the public not exceed 1 mSv/yr (100 mrem/yr), for continuous or frequent exposure from radiation sources other than medial exposures.

Most countries imposing limits on radiation from anthropogenic sources have endorsed the principle of apportionment of the total allowable dose. Many countries are in the process of endorsing a dose limit of 10 µSv/yr (1 mrem/yr) for LLRW that is 1 percent of the total allowable dose.

MULTIAGENCY RADIATION SURVEY AND SITE INVESTIGATION MANUAL

The *Multi-Agency Radiation Survey and Site Investigation Manual* (MARSSIM) provides a nationally consistent consensus approach to conducting radiation surveys and investigations at potentially contaminated sites. This approach is intended to be both scientifically rigorous and flexible enough to be applied to a diversity of site cleanup conditions. MARSSIM's title includes the term "survey" because it provides information on planning and conducting surveys and the term "site investigation" because the process outlined in the manual allows one to begin by investigating any site that may involve radioactive contamination.

The decommissioning that follows remediation requires a demonstration to the responsible federal or state agency that the cleanup effort was successful and that the release criterion (a specific regulatory limit) was met. This manual assists site personnel or others in performing or assessing such a demonstration. The demonstration of compliance involves three interrelated steps:

1. Translating the cleanup or release criterion (e.g., millisieverts per year, millirem per year, specific risk) into a corresponding derived contaminant concentration level (e.g., becquerels per kilogram or picocuries per gram in soil) through the use of environmental pathway modeling;
2. Acquiring scientifically sound and defensible site-specific data on the levels and distribution of residual contamination, as well as levels and distribution of radionuclides present as background, by employing suitable measurement techniques; and

3. Determining that the data obtained from sampling support the claim that the site meets the release criterion, within an acceptable degree of uncertainty, by applying a statistically based decision rule.

MARSSIM provides standardized and consistent approaches for planning, conducting, evaluating, and documenting environmental radiological surveys, with a specific focus on the final status surveys that are carried out to demonstrate compliance with cleanup regulations.

TABLE D-1 Exempt Quantities Established by Council Directive 96/29/EURATOM

Nuclide	Quantity (Bq)	Concentration (kBq/kg)	Nuclide	Quantity (Bq)	Concentration (kBq/kg)
H-3	10^9	10^6	Ga-72	10^5	10
Be-7	10^7	10^3	Ge-71	10^8	10^4
C-14	10^7	10^4	As-73	10^7	10^3
O-15	10^9	10^2	As-74	10^6	10
F-18	10^6	10	As-76	10^5	10^2
Na-22	10^5	10	As-77	10^6	10^3
Na-24	10^5	10	Se-75	10^6	10^2
Si-31	10^6	10^3	Br-82	10^6	10
P-32	10^5	10^3	Kr-74	10^9	10^2
P-33	10^8	10^5	Kr-76	10^9	10^2
S-35	10^8	10^5	Kr-77	10^9	10^2
Cl-36	10^6	10^4	Kr-79	10^5	10^3
Cl-38	10^5	10	Kr-81	10^7	10^4
Ar-37	10^8	10^6	Kr-83m	10^{12}	10^5
Ar-41	10^9	10^2	Kr-85	10^4	10^5
K-40	10^6	10^2	Kr-85m	10^{10}	10^3
K-42	10^6	10^2	Kr-87	10^9	10^2
K-43	10^6	10	Kr-88	10^9	10^2
Ca-45	10^7	10^4	Rb-86	10^5	10^2
Ca-47	10^6	10	Sr-85	10^6	10^2
Sc-46	10^6	10	Sr-85m	10^7	10^2
Sc-47	10^6	10^2	Sr-87m	10^6	10^2
Sc-48	10^5	10	Sr-89	10^6	10^3
V-48	10^5	10	Sr-90+	10^4	10^2
Cr-51	10^7	10^3	Sr-91	10^5	10
Mn-51	10^5	10	Sr-92	10^6	10
Mn-52	10^5	10	Y-90	10^5	10^3
Mn-52m	10^5	10	Y-91	10^6	10^3
Mn-53	10^9	10^4	Y-91m	10^6	10^2
Mn-54	10^6	10	Y-92	10^5	10^2
Mn-56	10^5	10	Y-93	10^5	10^2
Fe-52	10^6	10	Zr-93+	10^7	10^3
Fe-55	10^6	10^4	Zr-95	10^6	10

TABLE D-1 continued

Nuclide	Quantity (Bq)	Concentration (kBq/kg)	Nuclide	Quantity (Bq)	Concentration (kBq/kg)
Fe-59	10^6	10	Zr-97+	10^5	10
Co-55	10^6	10	Nb-93m	10^7	10^4
Co-56	10^5	10	Nb-94	10^6	10
Co-57	10^5	10^2	Nb-95	10^6	10
Co-58	10^6	10	Nb-97	10^6	10
Co-58m	10^7	10^4	Nb-98	10^5	10
Co-60	10^5	10	Mo-90	10^6	10
Co-60m	10^6	10^3	Mo-93	10^8	10^3
Co-61	10^6	10^2	Mo-99	10^6	10^2
Co-62m	10^5	10	Mo-101	10^6	10
Ni-59	10^8	10^4	Tc-96	10^6	10
Ni-63	10^8	10^5	Tc-96m	10^7	10^3
Ni-65	10^6	10	Tc-97	10^8	10^3
Cu-64	10^6	10^2	Tc-97m	10^7	10^3
Zn-65	10^6	10	Tc-99	10^7	10^4
Zn-69	10^6	10^4	Tc-99m	10^7	10^2
Zn-69m	10^6	10^2	Ru-97	10^7	10^2
Ru-103	10^6	10^2	Cs-134	10^4	10
Ru-105	10^6	10	Cs-134	10^7	10^4
Ru-106+	10^5	10^2	Cs-136	10^5	10
Rh-103m	10^8	10^4	Cs-137+	10^4	10
Rh-105	10^7	10^2	Cs-138	10^4	10
Pd-103	10^8	10^3	Ba-131	10^6	10^2
Pd-109	10^6	10^3	Ba-140+	10^5	10
Ag-105	10^6	10^2	La-140	10^5	10
Ag-108m+	10^6	10	Ce-139	10^6	10^2
Ag-110m	10^6	10	Ce-141	10^7	10^2
Ag-111	10^6	10^3	Ce-143	10^6	10^2
Cd-109	10^6	10^4	Ce-144+	10^5	10^2
Cd-115	10^6	10^2	Pr-142	10^5	10^2
Cd-115m	10^6	10^3	Pr-143	10^6	10^4
In-111	10^6	10^2	Nd-147	10^6	10^2
In-113m	10^6	10^2	Pm-147	10^7	10^4
In-114m	10^6	10^2	Pm-149	10^6	10^3
In-115m	10^6	10^2	Sm-151	10^8	10^2
Sn-113	10^7	10^3	Sm-153	10^6	10^2
Sn-125	10^6	10^2	Eu-152	10^6	10
Sb-124	10^6	10	Eu-152m	10^6	10^2
Sb-125	10^6	10^2	Eu-154	10^6	10
Te-123m	10^7	10^2	Eu-155	10^7	10^2
Te-125m	10^7	10^3	Gd-153	10^7	10^2
Te-127	10^6	10^3	Gd-159	10^6	10^3
Te-127m	10^7	10^3	Tb-160	10^6	10
Te-129	10^6	10^2	Dy-166	10^6	10^3

continues

TABLE D-1 continued

Nuclide	Quantity (Bq)	Concentration (kBq/kg)	Nuclide	Quantity (Bq)	Concentration (kBq/kg)
Te-129m	10^6	10^3	Ho-166	10^5	10^3
Te-131	10^5	10^2	Er-169	10^7	10^4
Te-131m	10^6	10	Er-171	10^6	10^2
Te-132	10^7	10^2	Tm-170	10^6	10^3
Te-133	10^5	10	Tm-171	10^8	10^4
Te-133m	10^5	10	Yb-175	10^7	10^3
Te-134	10^6	10	Lu-177	10^7	10^3
I-123	10^7	10	Hf-181	10^6	10
I-125	10^6	10	Ta-182	10^4	10
I-126	10^6	10^2	W-181	10^7	10^3
I-129	10^5	10^2	W-185	10^7	10^4
I-130	10^6	10	W-187	10^6	10^2
I-131	10^6	10^2	Re-186	10^6	10^3
I-132	10^5	10	Re-188	10^5	10^2
I-133	10^6	10	Os-185	10^6	10
I-135	10^6	10	Os-191	10^7	10^2
Xe-131m	10^4	10^4	Os-191m	10^7	10^3
Xe-133	10^4	10^3	Os-193	10^6	10^2
Xe-135	10^{10}	10^3	Ir-190	10^6	10
Cs-129	10^5	10^2	Ir-192	10^4	10
Cs-131	10^6	10^3	Ir-194	10^5	10^2
Cs-132	10^5	10	Pt-191	10^6	10^2
Cs-134m	10^5	10^3	Pt-193m	10^6	10^3
Pt-197	10^6	10^3	U-235+	10^4	10
Pt-197m	10^6	10^2	U-236	10^4	10^4
Au-198	10^6	10^2	U-237	10^6	10^2
Au-199	10^6	10^2	U-238+	10^4	10
Hg-197	10^7	10^2	U-238sec	10^3	1
Hg-197m	10^6	10^2	U-239	10^6	10^2
Hg-203	10^5	10^2	U-240	10^7	10^3
Tl-200	10^6	10	U-240+	10^6	10
Tl-201	10^6	10^2	Np-237+	10^3	1
Tl-202	10^6	10^2	Np-239	10^7	10^2
Tl-204	10^4	10^4	Np-240	10^6	10
Pb-203	10^6	10^2	Pu-234	10^7	10^2
Pb-210+	10^4	10	Pu-235	10^4	10^2
Pb-212+	10^5	10	Pu-236	10^4	10
Bi-206	10^5	10	Pu-237	10^7	10^3
Bi-207	10^6	10	Pu-238	10^4	1
Bi-210	10^6	10^3	Pu-239	10^4	1
Bi-212+	10^5	10	Pu-240	10^3	1
Po-203	10^6	10	Pu-241	10^5	10^2
Po-205	10^6	10	Pu-242	10^4	1
Po-207	10^6	10	Pu-243	10^7	10^3
Po-210	10^4	10	Pu-244	10^4	1
At-211	10^7	10^3	Am-241	10^4	10^3

TABLE D-1 continued

Nuclide	Quantity (Bq)	Concentration (kBq/kg)	Nuclide	Quantity (Bq)	Concentration (kBq/kg)
Rn-220+	10^7	10^4	Am-242	10^6	10^3
Rn-222+	10^8	10	Am-242m+	10^4	1
Rn-223+	10^5	10^2	Am-243+	10^3	1
Rn-224+	10^5	10	Cm-242	10^5	10^2
Rn-225	10^5	10^2	Cm-243	10^4	1
Rn-226+	10^4	10	Cm-244	10	10
Rn-227	10^6	10^2	Cm-245	10^4	1
Rn-228+	10^5	10	Cm-246	10^3	1
Ac-228	10^6	10	Cm-247	10^4	1
Th-226+	10^7	10^3	Cm-248	10^4	1
Th-227	10^4	10	Bk-249	10^6	10^3
Th-228+	10^4	1	Cf-246	10^6	10^3
Th-229+	10^3	1	Cf-248	10^4	10
Th-230	10^4	1	Cf-249	10^3	1
Th-231	10^7	10^3	Cf-250	10^4	10
Th-232sec	10^3	1	Cf-251	10^3	1
Th-234+	10^5	10^3	Cf-252	10^4	10
Pa-230	10^6	10	Cf-253	10^5	10^2
Pa-231	10^3	1	Cf-254	10^3	1
Pa-233	10^7	10^2	Es-253	10^5	10^2
U-230+	10^5	10	Es-254	10^4	10
U-231	10^7	10^2	Es-254m	10^4	10
U-232+	10^3	1	Fm-254	10^7	10^4
U-233	10^4	10	Fm-255	10^6	10^3
U-234	10^4	10			

SOURCE: EU (1996).

TABLE D-2 Derived USNRC Clearance Levels Assuming a 10 µSv/yr Primary Dose Standard (All Metals)

Mass Clearance Levels		Surficial Clearance Levels	
Radionuclide	NRC (Bq/g)	Radionuclide	NRC (Bq/cm^2)
H-3	2.E+04	H-3	2.E+04
C-14	6.E+02	C-14	7.E+02
Na-22	2.E–02	Na-22	3.E–02
P-32	8.E+01	P-32	9.E+01
S-35	1.E+03	S-35	2.E+03
Cl-36	4.E+00	Cl-36	5.E+00
K-40	2.E–01	K-40	3.E–01
Ca-41	8.E+01	Ca-41	1.E+02
Ca-45	1.E+02	Ca-45	2.E+02
Cr-51	4.E+00	Cr-51	5.E+00
Mn-54	1.E–01	Mn-54	1.E–01
Fe-55	1.E+04	Fe-55	1.E+04
Co-57	3.E+00	Co-57	3.E+00
Co-58	1.E–01	Co-58	1.E–01
Fe-59	9.E–02	Fe-59	1.E–01
Ni-59	2.E+04	Ni-59	3.E+04
Co-60	4.E–02	Co-60	5.E–02
Ni-63	8.E+03	Ni-63	1.E+04
Zn-65	5.E–02	Zn-65	6.E–02
Cu-67	5.E+00	Cu-67	6.E+00
Se-75	3.E–01	Se-75	4.E–01
Sr-85	2.E–01	Sr-85	2.E–01
Sr-89	9.E+01	Sr-89	1.E+02
Sr-90	1.E+00	Sr-90	1.E+00
Y-91	3.E+01	Y-91	3.E+01
Mo-93	7.E+02	Mo-93	9.E+02
Nb-93m	1.E+03	Nb-93m	2.E+03
Nb-94	6.E–02	Nb-94	7.E–02
Nb-95	1.E–01	Nb-95	2.E–01
Zr-95	1.E–01	Zr-95	2.E–01
Tc-99	5.E+01	Tc-99	7.E+01
Ru-103	2.E–01	Ru-103	3.E–01
Ru-106	5.E–01	Ru-106	6.E–01
Ag-108m	6.E–02	Ag-108m	7.E–02
Cd-109	2.E+01	Cd-109	3.E+01
Ag-110m	4.E–02	Ag-110m	4.E–02
Sb-124	6.E–02	Sb-124	6.E–02
I-125	4.E+00	I-125	5.E+00
Sb-125	2.E–01	Sb-125	3.E–01
I-129	2.E–01	I-129	2.E–01
I-131	4.E–01	I-131	5.E–01
Ba-133	4.E–01	Ba-133	4.E–01
Cs-134	2.E–02	Cs-134	2.E–02
Cs-137	4.E–02	Cs-137	5.E–02

TABLE D-2 continued

Mass Clearance Levels		Surficial Clearance Levels	
Radionuclide	NRC (Bq/g)	Radionuclide	NRC (Bq/cm^2)
Ce-141	4.E+00	Ce-141	4.E+00
Ce-144	3.E+00	Ce-144	4.E+00
Pm-147	9.E+02	Pm-147	1.E+03
Eu-152	9.E–02	Eu-152	1.E–01
Eu-154	8.E–02	Eu-154	1.E–01
Eu-155	9.E+00	Eu-155	1.E+01
Re-186	4.E+01	Re-186	5.E+01
Ir-192	8.E–02	Ir-192	1.E–01
Pb-210	7.E–02	Pb-210	9.E–02
Po-210	2.E–01	Po-210	2.E–01
Bi-210	3.E+02	Bi-210	4.E+02
Rn-222	1.E–01	Rn-222	1.E–01
Ra-223	6.E–01	Ra-223	6.E–01
Ra-224	2.E–01	Ra-224	2.E–01
Ac-225	7.E–01	Ac-225	8.E–01
Ra-225	6.E+00	Ra-225	7.E+00
Ra-226	6.E–02	Ra-226	7.E–02
Ac-227	3.E–02	Ac-227	4.E–02
Th-227	2.E+00	Th-227	2.E+00
Th-228	8.E–02	Th-228	9.E–02
Ra-228	1.E–01	Ra-228	1.E–01
Th-229	2.E–02	Th-229	3.E–02
Th-230	2.E–01	Th-230	2.E–01

E

Radiation Measurement

This appendix provides tutorial information about radioactivity, radiation, and their detection. It is important to understand the basic concepts of ionizing radiation, its interaction with matter, and its detection to be able to address many issues associated with the release of slightly radioactive solid material (SRSM) from regulatory control. Note that the levels of radioactive material concentration under consideration for release are very low relative to most licensed sources. In fact, these levels are close to those of the natural background radiation. As the concentration or amount of radioactive material decreases, detection and identification of the source or sources become more difficult.

First, consider some elementary but important aspects of matter. Atoms are composed of electrons that orbit around a nucleus. It is the number of electrons surrounding the nucleus that determines the chemical properties of the atom, and in an atom, the number of orbital electrons is equal to the number of protons in the nucleus, since protons are positively charged and electrons are negatively charged. Atoms gain electrons (to become anions), lose electrons (to become cations), or share electrons to form molecules. Neutrally charged particles—neutrons—also exist in the nucleus. The relative numbers of protons and neutrons play a key role in determining the stability of an atom's nucleus. Nuclei with the same number of protons but different numbers of neutrons are called isotopes.

Unstable nuclides—radionuclides—radiate particles and electromagnetic radiation when they transform to a more stable configuration. All isotopes of an element will behave the same chemically. For example, radioactive ^{60}Co will act just like stable ^{59}Co when steel is melted.

Radioactive material can be either naturally occurring or created by man. Radioactive decay is a random process. The half-life of a radionuclide is the

APPENDIX E 213

average time it takes for a sample of that radionuclide to reduce in quantity by one-half. The activity of a collection of radionuclides is a measure of the number of nuclear transformations per unit time occurring in a sample in units of becquerels (Bq) and curies (Ci). One becquerel is defined as one disintegrating nucleus per second. The curie is a customary unit that is equal to 3.7×10^{10} Bq. In any radiation measurement, there is a small statistical uncertainty resulting from the radioactive decay process.

It is the emitted radiation and its subsequent interaction with matter that can be detected. The type, energy, half-life, and frequency of detected radiation can be used to determine the amount of each radionuclide present in a sample. By comparing the quantity of each radionuclide present in a sample with the activity limits established from a dose standard, a determination can be made of whether the sample meets release criteria.

THE MEASUREMENT PROCESS

The method used to detect the radiation emitted from radioactive material plays an important role in determining the presence and quantity of a specific radionuclide or collection of radionuclides that are present. Two general approaches can be applied, each giving different levels of information. One method is to attempt to survey 100 percent of the material entering or leaving a facility. An example of this is the use of portal detectors to survey scrap metal entering a steel production site. The truck with a load of scrap pulls between two large detectors and slows or stops briefly while the load is "counted"; then, based on the number of counts obtained during the counting period, an essentially immediate determination is made of whether the load contains radioactive material. No attempt is made to identify or quantify the specific radionuclides that are present. An alternative method is to survey each piece of scrap metal individually, using a more sensitive detector capable of determining the identity and quantity of each of the radionuclides in the material by determining radiation type, energy, and activity. The first method has the clear advantage of being capable of a large throughput. Its major disadvantage is the inability to detect small quantities of radioactive material and its insensitivity to radiation that is easily stopped in matter. The second approach gives a very accurate and complete assessment of the radionuclide inventory (i.e., identity and quantity), but the process is tedious, leading to high personnel costs (more skilled personnel required) and low throughput. Thus, the measurement process selected will vary depending on the goal.

RADIATION TYPES AND INTERACTIONS

There are unique types and combinations of radiation emitted by individual radionuclides as they decay. This uniqueness permits identification of the radionuclide that decayed from its detected radiations. The most common types of radiation are alpha particles, beta particles, and gamma rays (or photons).

An alpha particle is a helium-4 nucleus with two protons, two neutrons, and a +2 charge. Alpha particles travel only a short distance before coming to a stop, having transferred all their kinetic energy to the target material. An alpha particle can usually be stopped by 2 to 3 cm of air or one sheet of paper. After the alpha particle stops, it simply picks up two free electrons and becomes a helium atom. Alpha particles are easy to shield and, thus, are of little hazard to humans when outside the body. Conversely, when alpha particles are emitted from radionuclides within the body, all of their kinetic energy is deposited in a small amount of tissue, resulting in a large, highly localized absorbed dose.

Beta particles originate in the nucleus when a neutron transforms to a proton. Beta particles are electrons that have been given this special name to differentiate them from the atomic orbital electrons. Like alpha particles, beta particles take energy away from the nucleus. Beta particles travel a longer distance through matter than alpha particles. A typical range of a beta particle is 1 to 3 meters in air or 0.1 to 1 cm in solids and liquids.

Radionuclides emit a third type of radiation, gamma rays, which are zero-mass, zero-charge photons. Usually, gamma photons are emitted in conjunction with particle decay to rid the nucleus of the remaining excess energy. Gamma photons also interact with a target material's orbital electrons, but with very low frequency compared to the interaction frequency of charged particles. This means that gamma photons are the most penetrating of the common types of radiation. The attenuation of photon radiation is described by an exponential relationship.

The interaction of radiation with matter is extremely important in the overall assessment of the radioactive material content of an unknown sample. To successfully measure the radioactive material in a sample, radiation emitted from the decaying nuclei must be able to penetrate everything between its point of emission and the detector. The radiation must then interact within the active volume of the detector.

Some radionuclides are difficult to measure because the radiation is not very penetrating. Radionuclides emitting only alpha or beta particles fall within this category. Special procedures must be used to quantify the radioactive material content of solid materials containing alpha- and beta-particle emitters. The difficulty in assaying materials contaminated with radionuclides that emit only particle radiation is getting the radiation to the detector.

Many radionuclides that decay by emission of alpha or beta radiation also simultaneously emit one or more gamma photons. Gamma photons are very penetrating relative to particles, with the exception of low-energy photons. For radioactive materials emitting gamma photons, different detectors (from those used for alpha and beta particles) are employed depending on the purpose of the measurement.

If it were necessary to determine only whether radiation is present, a detector that responds to alpha, beta, and gamma radiation would be preferred. An example of such a detector is the Geiger-Müller (GM) detector. A GM detector is a gas-filled chamber that is coupled to an electronic circuit to detect the pulses generated by a radiation interaction within the detector's active volume. These devices are portable and inexpensive. GM instruments are often used for initial surveys, since they register detected radiation events as "counts." By knowing the details of how the measurement was made and the sample characteristics, the radioactive material concentration in the sample can be estimated.

There are many other types of radiation detectors, including ion chambers, scintillation detectors, and solid-state detectors. Ionization chambers are air-filled detectors operated in the current mode. Ion chambers are insensitive at radiation intensities associated with the proposed clearance levels. Scintillation detectors are based on detection of the small light flashes produced by radiation interactions within a scintillation material. Scintillators can be manufactured in liquid, crystal, or plastic form. Because scintillators are usually designed to respond to one type of radiation, it is possible to eliminate some radionuclides from consideration when assaying an unknown sample. Additionally, the intensity of the flash is proportional to the energy; thus, scintillation detectors can be used to gain some information on the radiation's energy.

Solid-state detectors utilizing silicon or germanium are preferred for radiation spectroscopy because of the high-energy resolution possible from these devices. Solid-state detectors are available for particle and photon measurement. When coupled with a computer and spectral analysis software, these detectors provide a powerful tool for quantifying both the activity level and the radionuclide inventory in a sample.

It is perhaps easier to illustrate radiation detection and measurement procedures using two examples. The first example is the decision process made on scrap steel entering a steel plant. The objective of the measurement is to determine whether or not the shipment contains radioactive materials. A truckload of scrap is pulled between two detectors. If activity is detected, the shipment is rejected. Usually no attempt is made to sort the scrap or investigate the cause of the radiation alarm. Since the material is scrap metal contained in a truck, any particle radiation would be shielded from the detectors by the truck wall, the other scrap metal, and the air between the truck and the detectors. If sufficient quantities of radioactive materials that emit gamma rays are present, the detectors will respond accordingly.

This example illustrates several important points. The goal in many cases is to determine the presence or absence of radioactivity in a large amount of material. In order to maximize the probability of detection of the radiation from any radioactive materials present, the measurement system must be optimized, usually by the use of large gamma scintillation detectors. The go/no go type of system gives no information about the radionuclide inventory in the shipment, since the detectors used are not capable of providing sufficient data for radionuclide identification and the parameters necessary to convert from counts per unit time to activity are unknown.

A second hypothetical example is a U.S. Nuclear Regulatory Commission (USNRC) licensee who has a quantity of concrete for disposal that is probably not radioactive. However, the licensee is aware of the possibility that the concrete may have been irradiated with neutrons that would have created some radionuclides. External measurements with a survey instrument indicate that the activity, if present at all, is about at the background level. Thus, the problem is to determine whether the concrete contains neutron-produced radionuclides or only naturally occurring radionuclides. Since it would be reasonable to assume that neutrons could penetrate deeply into the concrete, it would follow that radionuclides could have been produced within the concrete, not just on its surface. An additional assumption would be that a wide variety of radionuclides could have been produced.

A solution would be to perform a measurement of the concrete in a laboratory. This requires collection of a statistically representative group of samples from the batch of concrete. Each sample would be analyzed carefully using standard methods to determine the radionuclides present and their respective activities. One method would be to crush the concrete to a fine powder and then count small volumes of the powder to eliminate source self-shielding, making it possible to determine if alpha or beta radiation is present. Spectroscopy could then be utilized to gather the data to determine the energies and intensities of each radiation type. Analysis of the data would yield a complete radionuclide inventory and determine whether any of the detected radionuclides were produced by neutron activation or whether they were naturally occurring.

This second example illustrates the difficulty with a quantitative assay of volumetrically contaminated or irradiated materials. Although exact activity inventory determinations are possible (and routinely performed), they utilize specialized, nonportable instrumentation in a laboratory environment. Such an analysis may take several weeks to complete at a fairly high cost (relative to simple scanning of materials). Thus, it is not realistic to anticipate that this type of analysis would be performed in most high-volume, high-throughput manufacturing processes.

BACKGROUND RADIATION

Background radiation is present in every counting situation. It results from several different sources, including naturally occurring radioactive materials, cosmic radiation, and man-made radionuclides from weapons tests. Some naturally occurring radionuclides have long half-lives, often more than a billion years. These are residual isotopes that were once present in much larger abundances but have slowly decayed with time. Examples of these include ^{40}K, ^{147}Sm, and ^{235}U. Other naturally occurring radionuclides are produced by activation by cosmic-ray bombardment of stable isotopes. An example of this is the production of radioactive ^{14}C from stable ^{14}N. Table E-1 gives some specific examples of background and man-made source activities. Since the distribution of radionuclides varies around the world depending on the geology of the area, some of these activities represent typical numbers. All detection systems must account for and subtract background levels to obtain true sample radioactive material concentrations.

TABLE E-1 Radiation Sources and Their Activities

Radiation Source	Radioactivity (Bq)
70 kg adult human (male) ^{40}K[a]	~5,000
1 kg of fresh vegetables[a]	10
1 kg of super phosphate fertilizer[b]	5,000
Air inside 2000 ft^2 home (radon) (593 m^3)[a]	36,000
Household smoke detector[b]	3,700-110,000
Radionuclide for medical diagnosis[c]	11-740 × 10^6
Radionuclide source for medical therapy[d]	3.7 × 10^{14}
1 kg natural uranium[a]	24 × 10^6
1 kg low-level radioactive waste (Class A, ^{137}Cs)[e]	4 × 10^7
1 kg of coal fly ash[b]	150-410
1 kg of granite (U, Th, K)[b]	72

[a]National Council on Radiation Protection and Measurements (NCRP) Report No. 94 (NCRP, 1987b).
[b]NCRP Report No. 95 (NCRP, 1987d).
[c]NCRP Report No. 100 (NCRP, 1989a).
[d]NCRP Report No. 105 (NCRP, 1989b).
[e]10 CFR Part 61.55

F

Stakeholder Reactions to the USNRC Issues Paper

This appendix describes alternative points of view expressed by a range of stakeholders responding to the U.S. Nuclear Regulatory Commission's (USNRC's) issues paper (64 Federal Register 35090-35100; June 30, 1999). The appendix is illustrative: it does not cover all groups with an opinion, nor does it cover all possible opinions (for this one should consult NUREG/CR-6682; USNRC, 2000d). In general, the committee found that positions taken by stakeholder groups on the alternatives presented in the USNRC's issues paper were often similar to those expressed when the below regulatory concern (BRC) policy was discussed 10 years earlier.

USNRC EFFORTS AT STAKEHOLDER INVOLVEMENT

Background

As the initial step in this process, the USNRC solicited comment on its June 30, 1999, issues paper (64 Federal Register 35090-35100; "Release of Solid Materials at Licensed Facilities: Issues Paper, Scoping Process for Environmental Issues and Notice of Public Meetings"), noting that it was the initial step in an

---

NOTE: Some of the displayed quotes in this appendix were recorded at the Rockville public meeting on May 9, 2000; others were from oral or written statements to the committee on March 26 and 27, 2001.

"enhanced participatory process" in which the USNRC would seek public input into its decision-making process (USNRC, 2000a). The envisioned participatory process would consist of various forums, invited written comments on the paper, and a Web site that contained the issues paper and other materials and invited public comment. The cornerstone of the enhanced process was four public meetings to provide stakeholder groups and the public an opportunity to learn about the USNRC's issues paper effort and respond to it.

The four sites for the meetings, held in 1999, included San Francisco, California (September 15-16); Atlanta, Georgia (October 5-6); Rockville, Maryland (November 1-2); and Chicago, Illinois (December 7-8). Despite numerous other attendees, public interest groups (such as environmental advocacy groups) did not attend the first two meetings. These groups provided a letter explaining why they would not attend. The last two meetings were attended by only a few of these interest group representatives; the others continued to boycott the public meetings. Although the USNRC had obtained more than 800 comment letters by December 2000, efforts by some groups to extend the comment period were denied by the USNRC (USNRC, 2000a). A public meeting in Rockville, Maryland, on May 9, 2000, was attended by a variety of public interest groups, including some that had boycotted earlier meetings with the USNRC (USNRC, 2000c). The May 2000 Rockville meeting between the USNRC and stakeholder groups was designed to "provide an opportunity to deal with a range of different people who have – reflect the diversity of views on this issue" (USNRC, 2000c, p. 2). Three panels were headed by the chairman of the USNRC and were conducted with some opportunity for presenters to enter a dialogue with commissioners. The summary of the public meetings and written comments (USNRC, 2000d) and the proceedings from the Rockville stakeholder-Commission meeting (USNRC, 2000c) form two of four databases for this appendix. The other two include the summary of public comments at the four public meetings, NUREG/CR-6682 (USNRC, 2000d), just prior to the May 9, 2000, Rockville dialogue and various presentations made to this committee by different stakeholder groups including nuclear industry groups, agreement states, environmental interest groups, and others from the metals and concrete industries.

The USNRC (2000a) staff report, SECY-00-0070, and the ICF Kaiser report (USNRC, 2000d) are the two major summaries of all 900 (written) comments as well as additional oral comments. Both efforts adequately provide the reader with the range of responses to the issues paper categories, but they do not offer a sense of the intensity of the views expressed. In addition, both SECY-00-0070 and the ICF Kaiser report tie comments and analyses back to the preexisting issues paper categories (process alternatives and technical approach categories). The difficulty of adequately summarizing and analyzing these comments (without some sort of weighting, content analysis, and/or statistical analysis) is clearly evident in the documents.

Approach

The approach adopted in this report deviates from these two prior efforts in that it identifies stakeholder opinions without deliberate attempt to tie the opinions back to categories identified in the issues paper. As a result, some opinions correlate well with issues paper categories and some do not. The rationale for this approach is to avoid the misperception that all comments made correlate perfectly with categories identified in the issues paper; clearly, some stakeholders' opinions contravene issues-paper-defined categories, while others embody categories not originally envisioned by the issues paper (for a review of the stakeholder positions relative to issues paper categories, refer to Chapter 8, Table 8-1).

The approach utilized is a qualitative analysis of data obtained from presentations made directly to the study committee (see Appendix B for a complete list of presenters) and from the Rockville meeting. The perspectives that developed from the committee's analysis were then confirmed by reviewing both the ICF summary report and the USNRC staff report to the commissioners summarizing the input from all of the hearings.

The committee's approach was to extract remarks and aggregate them into categories that demonstrate gradations of opinion. These gradations ranged from no clearance to support for a specific release standard. The gradations identified are the following:

- Cannot support release (clearance) for recycle or disposal;
- Cannot engage in a dialogue with the USNRC because the dialogue process is itself tainted;
- Recommend delaying a decision on whether to conduct a rulemaking until public comments can be integrated into the USNRC's decision framework;
- Recommend restricted release (conditional clearance);
- Recommend continuing case-by-case, but with uniform national dose-based criteria; and
- Recommend setting a specific clearance standard, but with some exceptions for special groups such as the metals recycling industry.

Each of these gradations is discussed further below. In addition, options are discussed that fall outside those originally envisioned by the USNRC, which nonetheless need to be identified and considered.

Stakeholder Positions

Cannot Support Release for Recycle or Disposal

Illustrative stakeholders in this category consisted of environmental public interest groups (Nuclear Information and Resource Service, Public Citizen, New

APPENDIX F *221*

England Coalition on Nuclear Pollution), which generally believe that the solid materials should be "regulated, monitored, and isolated from general commerce."[1] These stakeholders tended to share the following perceptions:

- The U.S. Nuclear Regulatory Commission already has a position (i.e., a free release standard that it hopes to promulgate), and the public involvement process is just that—a process, not a meaningful dialogue.
- Multiple and synergistic effects are possible from a release that is recycled into numerous sources for public use, and these effects have not been well characterized by the USNRC or any scientific body.
- The USNRC consistently uses so-called scientific evidence to justify the bases for its decisions, even though reasonable people might (and often do) disagree as to whether these bases can be substantiated.
- Releases of radioactive materials cannot be tracked or otherwise controlled, thereby raising doubts about the role they could play in the stream of commerce not only during their initial use but also during any subsequent uses.
- The USNRC role in developing a standard is self-perpetuating in that the USNRC is attempting to justify its position by "punting"[2] to an international standard that the USNRC itself had a role in creating.
- The fact that some radioactive materials already exist in the stream of commerce (whether natural or man-made) is not sufficient to justify additional releases.
- The USNRC's true intent is economic, that is, to enable recycling of large amounts of contaminated material, which will benefit no one but the nuclear industry.
- The USNRC should seriously consider a "no-release" option; however, no group stated a specific, preferred process or technical alternative for how the materials should be treated, other than to request development of a scenario by which the materials would not be released.

Two observations are offered based on these comments:

- Stakeholder viewpoints reflect an overall lack of trust in the USNRC.
- Since most stakeholders assumed that the USNRC's true objective is to recycle the material, they were taken aback when asked whether removing recycling from the equation would make a difference. Most indicated that it would, in fact, make a significant difference in the degree of their opposition to the proposed rulemaking.

---

[1] Oral testimony of Dianne D'Arrigo, Nuclear Information and Resource Service, Rockville public meeting. May 9, 2000.

[2] Oral testimony of Wenonah Hauter, Public Citizen, Rockville public meeting. November, 1999.

"Your job is to prevent exposures to the public and the environment—not convince us that it's a trivial amount." —*Wenonah Hauter, Public Citizen*

"Informed and concerned citizens may choose to remove themselves from the vicinity of nuclear facilities. No such information and no such choice exists when contaminated materials are free-released for recycle."
—*New England Coalition on Nuclear Power*

Cannot Engage in a Dialogue with the USNRC Because the Dialogue Process Is Itself Tainted

Illustrative stakeholders in this category consisted of a public interest group, Public Citizen, and the Allied Industrial Chemical and Energy Workers Union. These stakeholders shared most or all of the concerns noted above and additionally rejected the process and technical framework set forth by the USNRC. These stakeholders tended to perceive the following:

- The USNRC is asking the wrong question and may thus be avoiding discussion of all possible options and alternatives. To these stakeholders, the question is not, Should we conduct a rulemaking—why or why not? Rather, it is, Can we have an honest dialogue that would enable consideration of all the options and alternatives—even those options and alternatives that the USNRC dislikes?
- The USNRC and the National Academy of Sciences (NAS) through the National Research Council both mistakenly believe the primary issues to be technical issues involving formulation, in the abstract, of some notional dose that ensures public health and safety, whereas the real issue is that the USNRC has an "empirical record of institutional incompetence"[3] that shows a startling incapacity to technically abide by public protection standards."[4]
- The real task is for the NAS to undertake a thorough public review of whether the Department of Energy (DOE) and its contractors cannot reasonably be relied on by the public to abide by whatever radiation protection standards might, in the abstract, appear reasonable.
- Although the USNRC does not regulate DOE facilities, any standard adopted by the USNRC would in fact be implemented by DOE, since DOE is perceived to be the prime promoter of the unrestricted release of contaminated metals.[5]

---

[3]Oral remarks by Dan Guttman to the committee, March 27, 2001.

[4]Written testimony of Dan Guttman, presented to the committee March 27, 2001, p. 3.

[5]Ibid, p. 17. This theme was also reflected in testimonies to the committee by other public interest groups.

Two observations emerge from these comments:

1. The groups in this category do not believe that it is possible to engage in a dialogue until other issues of whether or not the USNRC can reliably protect public health and safety are resolved.
2. Again, since most stakeholders assumed that the USNRC's true objective is to recycle the material, they were taken aback when asked whether removing recycling from the equation would make a difference. Most indicated that it would, in fact, make a significant difference in the degree of their opposition to the proposed rulemaking.

"Why did the NAS—an entity with essential responsibility for vouchsafing the integrity of scientific advice to the nation—accept without any evident question a framework for its work which it knew, or should have known, effectively excluded from inquiry most important facts bearing on the protection of the public?"
—Dan Guttman

Recommend Delaying a Decision on Whether to Conduct a Rulemaking Until Public Comments Can Be Integrated into the USNRC's Decision Framework

Illustrative stakeholders in this category include groups as diverse as the scrap recycling industry, the Natural Resources Defense Council (NRDC), the Association of Radioactive Metals Recyclers (ARMR), and the Association of State and Territorial Solid Waste Management Officials (ASTWMO). They believe that the question of whether to conduct a rulemaking should be delayed in order to obtain substantive representation from all the affected stakeholders—that is, to incorporate the stakeholders' viewpoints into the decision framework. Stakeholders in this category generally tended to perceive the following:

- While it is possible to arrive at a defensible, scientific standard, the thought of radioactive materials entering the recycling stream elicits strong fears and concerns on the part of the public.
- The USNRC's investigation should focus not only on the technical issues but also on understanding and integrating public concerns into the overall process.
- As the NRDC suggested, the USNRC is unwilling or unable to explain the basis for its position, and fundamental questions should be answered as to (1) why contaminated solid materials had to be recycled in the first place and (2) how the USNRC would propose to regulate these materials in a way that protects public health and safety.

Specific suggestions were also made, as follows:

- ASTWMO[6] suggested that the USNRC itself might not have explored all the consequences of a rulemaking such as the following:
  —Would rulemaking consume fewer resources than continuing to use case-by-case?
  —Would the increased credibility of the USNRC resulting from delaying the process be of more benefit than making a rule?
  —How important is the rule to licensees?
  —What are the economics of the problem?
- ARMR[6] suggested that a demonstration plan acceptable to both industry and the public should be developed; this plan should be the collaborative work of key stakeholders to gain their acceptance for determining impacts (e.g., to an industry).
- ARMR[6] suggested that the appropriate next step in the USNRC's process would be to convene a balanced stakeholder committee that would report to the USNRC and would provide it with criteria for acceptable release, recycle, and reuse.
- These comments suggest that the stakeholders generally felt that no step in the USNRC's public outreach process had thus far been able to represent and integrate stakeholder concerns into the decision framework.

"It's not that the public doesn't understand—it's just that they have a different perspective based on risk and government credibility."
—*Jeff Deckler, Department of Human Health and Environment for the State of Colorado, representing ASTSWMO*

"If you were going into end uses that were very clear and controlled, and we had confidence in how the material was being surveyed and how measurements were being made, what you're proposing is something we might consider."
—*Natural Resources Defense Council*

Recommend Restricted Release (Conditional Clearance)

Illustrative stakeholders in this category included the metals and concrete industries. Both expressed serious concerns about the potential economic damage to their markets from free release. Both support a restricted use concept, in which

---

[6]The only group on record that took this position.

solid waste re-use would be limited to selected purposes and subject to a high degree of control.

The concrete industry (National Ready Mixed Concrete Association) generally perceives the following:

- Unrestricted release would force both the ready-mixed concrete producer and the consumer to assume liability or cost for potentially contaminated materials.
- Unrestricted use of contaminated materials could put an extreme burden on unqualified handlers of radioactive materials, such as ready-mixed concrete producers.
- It would be difficult to conceive of unrestricted use of contaminated concrete, since recycled concrete—whether contaminated or not—does not have the best record in the construction industry.
- Unrestricted widespread use of any of the solid materials from licensed facilities is unacceptable.
- Restricted use should be defined to include only single point users where contact for exposure of the general population is minimal; examples could include non-water supply concrete dams for flood control, deep concrete foundations, or concrete containment facilities used as licensed storage facilities.
- Restricted use should entail licensing these facilities as low-level waste facilities.

Illustrative stakeholders in the metals industry included the Steel Manufacturers Association, the American Iron and Steel Institute, and the Metals Industry Recycling Coalition. They generally perceived the following:

- Radioactively contaminated scrap has no value and could in fact contribute to economic losses for scrap recyclers, since free release could damage the market for steel products by eroding public confidence in the safety of steel products.
- Free release could also add substantially to costs by forcing steel mills to go to extremes to protect against volumetrically contaminated materials that could cause a radioactive melt; recycling is viewed by the industry as a way for DOE to shift responsibility to the mills; and if sensor alarms go off too frequently, they may be ignored by employees—even if the alarm is truly warranted.
- No unrestricted release of any contaminated radioactive steel or other metals should be permitted from USNRC-licensed facilities, even if the steel meets dose-based release levels that the committee might recommend and the USNRC adopt.

- Material should be reused by DOE, stored or disposed on-site at the licensed facility, or disposed of off-site.
- Products from a licensed facility that are to be used for their original purpose off-site could be released without special restrictions if they meet a dose-based standard; those not used for their original purpose could be released to landfills or for dedicated nuclear-related uses such as at USNRC-licensed or DOE facilities.

The following observations were made by spokespeople for the metals and concrete industries: both industries made a useful distinction between recycling and disposal, and it is recycling that poses the perceived economic threat to them.

"The last thing the [metals] industry needs is to have a release standard that allows thousands or potentially millions of tons of steel that will meet the release standard but exceed our detectors coming into the mills. It will essentially shut down our ability to control for orphan sources."
—*John Wittenborn, Metals Industry Recycling Coalition*

"Faced with the challenges of closing licensed facilities and handling contaminated concrete, it is logical to conclude that a rule regarding release of contaminated materials from licensed sites should be made. It is not, however, an adequate conclusion in our opinion that these materials should be placed in unrestricted use or even restricted use without further definition. Concrete, as several other construction materials, is ubiquitous to our society. The concept of concrete framed buildings across the United States being made with radioactive materials housing millions of people exposing them to potential radioactive material greater than background exposure is contrary to the charter of the NRC."
—*Robert A. Garbini, National Ready Mixed Concrete Association*

Recommend Continuing Case by Case, but with Uniform National Dose-Based Criteria

Several individual states and the Organization of Agreement States (OAS) gave the committee information regarding their views and activities related to clearance of slightly radioactive solid material (SRSM). The OAS recommended the development of standards that would apply nationally and felt that the standards should address "free release" of material for unrestricted use. It commented that the approach should be "similar to the USNRC's tiered approach for license termination." It was suggested that consideration be extended to radioactive materials generated from technologically enhanced naturally occurring radioactive

APPENDIX F 227

material (TENORM) and naturally occurring and accelerator-produced radioactive material (NARM) sources.

In general, states have been applying case-by-case decisions to radioactive materials that are considered for alternative disposal, reuse, recycle, or clearance from the regulatory process. They have done so under their agreement states' authority and existing regulations. It seemed clear to the committee that while this process has been ongoing, a more formal and uniform process would be desirable.

Illustrative stakeholders in this group were the Conference of Radiation Control Program Directors (CRCPD) and the Organization of Agreement States,[7] which suggested continuing the case-by-case approach but using uniform, national dose-based criteria. The CRCPD and OAS see the main limiting factor under the current case-by-case approach as licensees' using different survey equipment with different detection limits, leading to inconsistencies in the overall approach. The CRCPD and OAS position suggested that states want a more consistent application of criteria, as well as uniform criteria. They proposed that because a value of 1 mrem/yr is not only a trivial dose but also the basis for the American National Standards Institute (ANSI) standard, it readily suggests itself as an easy common denominator.

Recommend Setting a Specific Clearance Standard, but with Some Exceptions for Special Groups Such as the Metals Recycling Industry

Illustrative stakeholders in this category included the Health Physics Society, the Nuclear Energy Institute, the American Nuclear Society's Special Committee on Site Restoration and Cleanup Standards, and the CRCPD E-23 Committee on Resource Recovery and Radioactivity. These stakeholders generally shared in the following perceptions:

- Lack of a consistent acceptance criteria provides inconsistent public protection, undermines public confidence, wastes resources, and perpetuates liability.
- In the absence of a clearance standard, there may be some wastage of potentially recyclable materials.
- Regulatory Guide 1.86 (AEC, 1974) contains surface contamination guidelines only (no volumetric criteria) and is not dose based.
- Current regulations are inconsistently applied.
- Current regulations do not cover recycling.

---

[7]Testimony of Steve Collins, Illinois Department of Nuclear Safety, representing both organizations, at USNRC, May 9, 2000.

- Current regulations are inconsistent with the standards adopted by the international community.
- A national clearance standard should be developed through rulemaking and should embrace ANSI N13.12 because it is a consensus standard, uses the same dose criteria as the International Atomic Energy Agency, uses practical screening values, can be verified with available instruments, and would establish a "floor." To these ends, the standard should be expedited for direct reuse and direct disposal.
- The steel recycling industry deserves special consideration because orphaned sources are a risk to public health, steel workers, and the steel industry.[8]
- There is a need to distinguish "disposal" from "recycle."
- The following observation is based on the points above; even those stakeholders who essentially support the development of a specific clearance standard would argue for special consideration to be given to the metals industry.

"We continue to advocate for the eventual promulgation of clear, consistent and enforceable regulations based upon a one millirem annual dose criterion and nuclide specific concentration guidelines."
—*Kathleen McAllister, Committee on Resource Recovery and Radioactivity*

Options Beyond Those Originally Envisioned by the USNRC Have to Be Identified and Considered in Any Further Stakeholder Involvement Process.

As can be seen from the matrix in Table 8-1, all stakeholder opinions do not neatly line up with process and technical alternatives initially envisioned by the USNRC in its issues paper, notably the section of the matrix that refers to "other" alternatives. This category includes the following:

- Groups who felt strongly that there should be no release but were not prepared to formulate specific no-release scenarios;
- Groups who essentially supported a rulemaking but who felt that the rulemaking should be delayed until all public comments have been integrated into the USNRC's decision framework; and

---

[8]The special exception for metals recyclers (and others such as the concrete industry) was not uniformly shared. Written testimony of Kathleen McAllister, chair, CRCPD E-23 Committee on Resource Recovery and Radioactivity, to the committee on March 27, 2001: "Despite inconveniences caused to them. . . . [i]t is reasonable to assume that landfills and scrap recycling yards, as well as municipal public sewer facilities, and possibly concrete facilities will take it upon themselves to install radiation monitoring equipment . . . ." p. 3.

APPENDIX F 229

- Groups who were unwilling to engage in discussion of "new" issues surrounding the release of solid materials until the "old" issues involving lack of public trust and confidence in the USNRC's ability to protect the public can be resolved.

The USNRC had expected to receive comments on the issues paper that would offer new options and alternatives. In this light, the discussion of stakeholder views above and the matrix of options in Chapter 8 may be of some value in framing other options and alternatives.

G

Acronyms and Glossary

ACRONYMS

AEA	Atomic Energy Act of 1948, as amended in 1954
AEC	U.S. Atomic Energy Commission
ALARA	as low as is reasonably achievable
ANS	American Nuclear Society
ANSI	American National Standards Institute
ARMR	Association of Radioactive Metals Recyclers
ASTSWMO	Association of State and Territorial Solid Waste Management Officials
BEIR	Committee on the Biological Effects of Ionizing Radiation
BRC	below regulatory concern
BSS	Basic Safety Standards (EC)
BWR	boiling water reactor
CERCLA	Comprehensive Environmental Response, Compensation, and Liability Act
CFR	U.S. Code of Federal Regulations
CNWRA	Center for Nuclear Waste Regulatory Analyses
CRCPD	Conference of Radiation Control Program Directors
DCGL	derived concentration guideline level
DoD	U.S. Department of Defense
DOE	U.S. Department of Energy
dpm	disintegrations per minute
DU	depleted uranium

EC	European Commission
EMC	elevated measurement comparison
EPA	U.S. Environmental Protection Agency
ERDA	U.S. Energy Research and Development Administration
EU	European Union
GM	Geiger-Müller
GSD	geometric standard deviation
HPGe	high-purity germanium
HPS	Health Physics Society
IAEA	International Atomic Energy Agency
ICRP	International Commission on Radiological Protection
IE	Office of Inspection and Enforcement (USNRC)
INSC	International Nuclear Societies Council
ISFSI	independent spent fuel storage installation
LLRW	low-level radioactive waste
LLWPAA	Low Level Radioactive Waste Policy Amendments Act of 1985
MARSSIM	*Multi-Agency Radiation Survey and Site Investigation Manual*
MDC	minimum detectable concentration
NARM	naturally occurring and accelerator-produced radioactive material
NAS	National Academy of Sciences (U.S.)
NCRP	National Council on Radiation Protection and Measurements
NEPA	National Environmental Policy Act
NESHAP	National Emission Standards for Hazardous Air Pollutants
NORM	naturally occurring radioactive material
NPL	National Priorities List
NRC	National Research Council
NRDC	Natural Resources Defense Council
OAS	Organization of Agreement States
ppm	part per million
PRA	probabilistic risk assessment
PWR	pressurized water reactor
RCRA	Resource Conservation and Recovery Act
SAIC	Science Applications International Corporation
SCA	Sanford Cohen & Associates, Inc.
SDMP	Site Decommissioning Management Plan
SI	international system of units
SRSM	slightly radioactive solid material

TEDE	total effective dose equivalent
TENORM	technologically enhanced naturally occurring radioactive materials
TSD	technical support document
UNSCEAR	United Nations Scientific Committee on the Effects of Atomic Radiation
USACE	U.S. Army Corps of Engineers
USNRC	U.S. Nuclear Regulatory Commission

GLOSSARY

agreement state — Section 274 of the AEA authorizes the Commission to enter into an effective agreement with the governor of a state to allow that state to assume the USNRC's authority to regulate certain types of materials licensees only. Reactor licensees remain the exclusive domain of the USNRC. Today there are 32 agreement states, which have implemented state regulations that are equivalent and compatible with the USNRC's regulations, as required by section 274(d) of the AEA. The materials licensees that a state can regulate include those that use or possess source material, byproduct material or special nuclear material in quantities not sufficient to form a critical mass (less than 350 grams for uranium-235).

de minimis — Shortened form of *de minimis non curat lex*, which is Latin for the common law doctrine stating, in free translation, that "the law does not concern itself with trifles." A *de minimis* amount of something (e.g., a dose) is one at or below which statutory or regulatory controls on larger amounts would not apply.

11(e)2 materials — Materials defined in section 11(e)(2) of the Atomic Energy Act (AEA) of 1954, as amended to be the tailings or waste produced by the concentration or extraction of uranium or thorium ore processed primarily for its source content. This definition was added in a 1978 by section 201 of the Uranium Mill Tailings Radiation Control Act, which amended the AEA.